Thomas Radford

Observations on the Caesarean Section, Craniotomy, and on Other Obstetric Operations

With cases. Second Edition

Thomas Radford

Observations on the Caesarean Section, Craniotomy, and on Other Obstetric Operations
With cases. Second Edition

ISBN/EAN: 9783337254124

Printed in Europe, USA, Canada, Australia, Japan

Cover: Foto ©berggeist007 / pixelio.de

More available books at **www.hansebooks.com**

ON THE CÆSAREAN SECTION,

CRANIOTOMY, &c.

OBSERVATIONS

ON THE

CÆSAREAN SECTION,

CRANIOTOMY,

AND ON

OTHER OBSTETRIC OPERATIONS.

With Cases.

BY

THOMAS RADFORD, M.D.

F.R.C.P. EDIN., F.R.C.S. ENG., ETC.

HONORARY CONSULTING PHYSICIAN TO ST. MARY'S HOSPITAL, AND THE
MANCHESTER AND SALFORD LYING-IN HOSPITAL.

SECOND EDITION.

LONDON :

J. & A. CHURCHILL, NEW BURLINGTON STREET.

1880.

CONTENTS.

PART I.

PART II.

Introductory Remarks   75

On Craniotomy and Cæsarean Section .   .   .   81

On the Statistics as shown in the Cases contained in Table II. 96

On the Maternal Mortality in Tables II. and III. . 97

On Infantile Mortality   ,,   ,,   . 99

On Infants Saved   .   . 101

On Exhaustion   101

On Peritonitis .   .   102

On Hæmorrhage .   .   104

On the Successful Cases .   .   .   106

On the Administration of Chloroform   108

On the Use of Ether-Spray   .   110

On the Use of Sutures 111

On the Operation   .   .   . 113

On the Instruments Used in Craniotomy   120

On the Use of the Perforator   120

    ,,   ,,   Crotchet   .   .   120

    ,,   ,,   Craniotomy Forceps .   121

    ,,   ,,   Osteotomist .   .   121

    ,,   ,,   Cephalotribe .   123

    ,,   ,,   Cranioclast .   123

    ,,   ,,   Vertebral Hook .   125

    ,,   ,,   Wire Ecraseur   125

    ,,   ,,   Basilyst .   .   .   126

EXPLANATION OF PLATES.

Explanation of Figure 1st.

This sketch accurately represents the state and dimensions of the brim of the pelvis belonging to Mary Ashworth (Case 27).

A, B.　A line from the sacro-iliac junction, to the linea ileo-pectinea opposite, on the right side, six-eighths of an inch.

C, D.　Ditto, ditto, on the left side, five-eighths of an inch.

E, F.　A line from the upper edge of the fifth lumbar vertebra, to the linea ileo-pectinea behind the acetabulum, on the right side, three-quarters of an inch.

G, H.　Ditto, ditto, on the left side, half an inch.

I, K.　Line across the anterior slit, two-eighths of an inch.

Explanation of Figure 2nd.

This outline shows the figure and dimension of the brim of the pelvis belonging to Mary Nixon (Case 30), as accurately as could be ascertained by an examination *per vaginam.*

A, B.　A line from the linea ileo-pectinea behind the acetabulum to the most projecting point on the back part of the pelvis, on the right side, nearly one inch.

C, D.　Ditto, ditto, on the left side, three-quarters of an inch.

Explanation of Figure 3rd.

This sketch accurately represents the brim of the pelvis belonging to Mary Forrest (Case 41).

A, B.　From the upper edge of the fifth lumbar vertebra to the ilium just behind the acetabulum, on the right side, five-eighths of an inch.

C, D.　From the same points on the left side, an inch and one-eighth.

E, F.　From the inside of symphysis pubis to the upper part of the fifth lumbar vertebra, two inches and six-eighths.

G, H.　Across the anterior slit, half an inch.

Explanation of Figure 4th.

The outline in this figure indicates the shape and dimensions of the brim of the pelvis of Mrs. Sankey (Case 49), drawn from the measurements obtained by an accurate examination *per vaginam*, previous to the Cæsarean section.

A, B. A line from the sacro-iliac junction to the iliac-pectineal line opposite, on the right side, one inch and a half.

C, D. Ditto, ditto, on the left side, nearly the same.

E, F. A line from the most projecting part behind, to the pubis near the jutting forward of these bones, three-quarters of an inch on right side.

G, II. Ditto, ditto, on the left side, three-quarters of an inch.

I, K. Space between the jutting forward of the pubes, half an inch.

Explanation of Figure 5th.

Sketch of the same (Mrs. Sankey's) pelvis, taken from the diagrams and measurements given in *The British Record*, which were obtained at the post-mortem examination.

A, B, C, D. Spaces between the acetabula and promontory of the sacrum, on each side of the pelvis, one inch and a quarter.

E, F, G, II. The narrowest measurement, one-half to three-quarters of an inch.

It is a misfortune that a more minute and accurate account of the pelvis of this woman was not obtained. It was evidently a little more contracted at her death than it was at the time of the Cæsarean section.

Explanation of Figure 6th.

This sketch shows " a full-sized outline of the brim of the late Mrs. Toft's pelvis, taken from a section of the cast" (Case 51).

Mr. Braid states that the points of two fingers laid side by side could not pass the brim, " excepting about half an inch exactly opposite the symphysis pubis, and there the fingers had barely room to pass. Beyond this, on either side, there was very little more than an inch of available space in the antero-posterior direction."

Explanation of Figure 7th.

The outline in this figure indicates the shape and dimensions of the brim of the pelvis of Mrs. Haigh (Case 53), drawn from the measurements obtained by an accurate vaginal examination previous to the Cæsarean section.

A, B, C, D. The widest part of brim, on each side, an inch and one-eighth.

E, F, G, H. The shortest measurements in the conjugate diameter, seven-eighths of an inch.

I, K. Across the anterior slit, half an inch.

Explanation of Figure 8th.

This sketch shows the extremely contracted state of the brim of the pelvis of the same woman, as found at her death (Case 53, sequel).

A, B. Widest part of brim on the right side, two-eighths of an inch.

C, D. Ditto, ditto, on the left side, rather less.

E, F. Narrowest part on right side, one-eighth of an inch.

G, H. Ditto, ditto, on left side, less.

I, K. Widest space in anterior slit, two-eighths of an inch.

The pubes at the commencement of the jutting forward, touch each other.

A ball whose diameter is two-eighths of an inch will only just pass through the opening between the projecting forward of the lumbar vertebræ and the ossa pubis.

Explanation of Figure 9th.

This sketch accurately represents the brim of Ann Kenion's pelvis (Case 58).

A, B. A line from the sacro-iliac junction to the linea ileo-pect., just behind the posterior edge of the acetabulum, on the right side, an inch and two-eighths.

C, D. Ditto, ditto, on the left side, an inch and a half.

E, F. A line from behind the anterior part of the acetabulum, to the upper edge of the fifth lumbar vertebra, on the right side, six-eighths of an inch.

G, H. Ditto, ditto, on the left side, an inch and one-eighth.

I, K. Across the middle part of the anterior slit, nearly half an inch.

L, M. Across the widest part, ditto, six-eighths of an inch.

From the line, *L, M,* to the lower portion of the fourth lumbar vertebra, an inch and one-eighth.

Explanation of Figure 10th.

A view of the superior aperture of Elizabeth Sherwood's pelvis, as described by Dr. Osborn, taken from Dr. Hull.

A, B. The dimension of the widest space, from before to behind, on the left side of the pelvis, three-quarters of an inch.

C, D. Dimension of the space betwixt the pubes and the sacrum, three-quarters of an inch.

E, F. Dimension of the widest space, from before to behind, in the right side of the pelvis, one inch and three-quarters.

G, H. The dimension of the space, betwixt the right os ilium and the os sacrum, two inches and a quarter.

Explanation of Figure 11*th.*

A, B. From the middle of the fifth lumbar vertebra to the ilium, just behind the acetabulum on the right side, one inch.

G, H. The same on the left side, one inch and a half.

E, F. From the upper and fore part of the fifth lumbar vertebra to the acetabular part of the pubis on the right side, half an inch.

C, D. The same on the left side, one inch and a quarter.

I, K. From the inside of the symphysis pubis to the upper edge of the fifth lumbar vertebra, two inches.

L, M. Across the anterior projection of the pubes, one inch and one-eighth.

*** This sketch represents the brim of the pelvis of a woman whose case is numbered 114 in Table III.

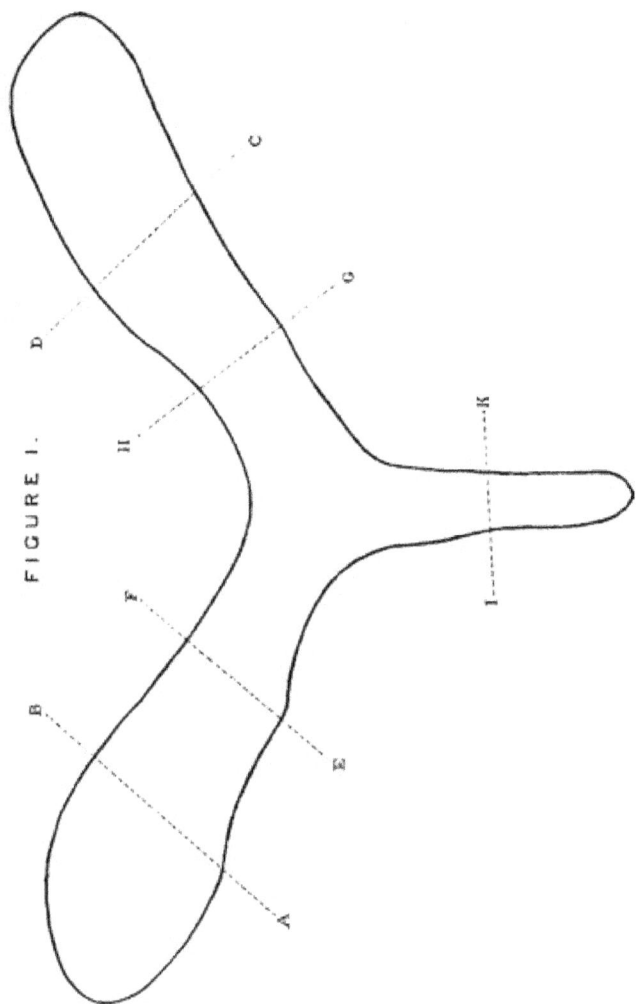

FIGURE 1.

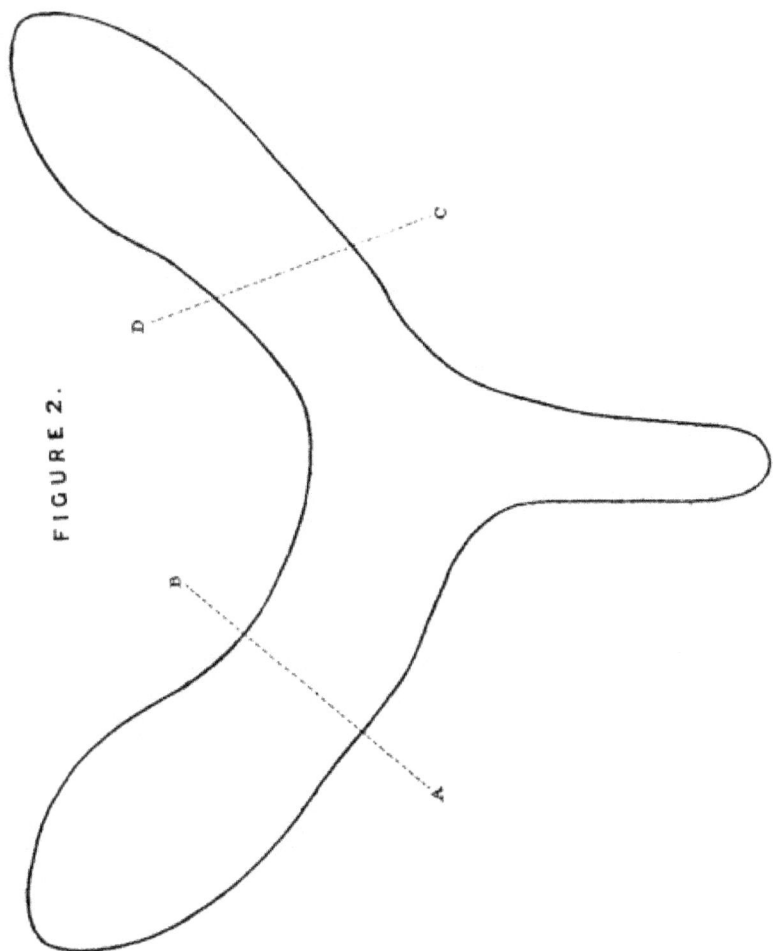

FIGURE 2.

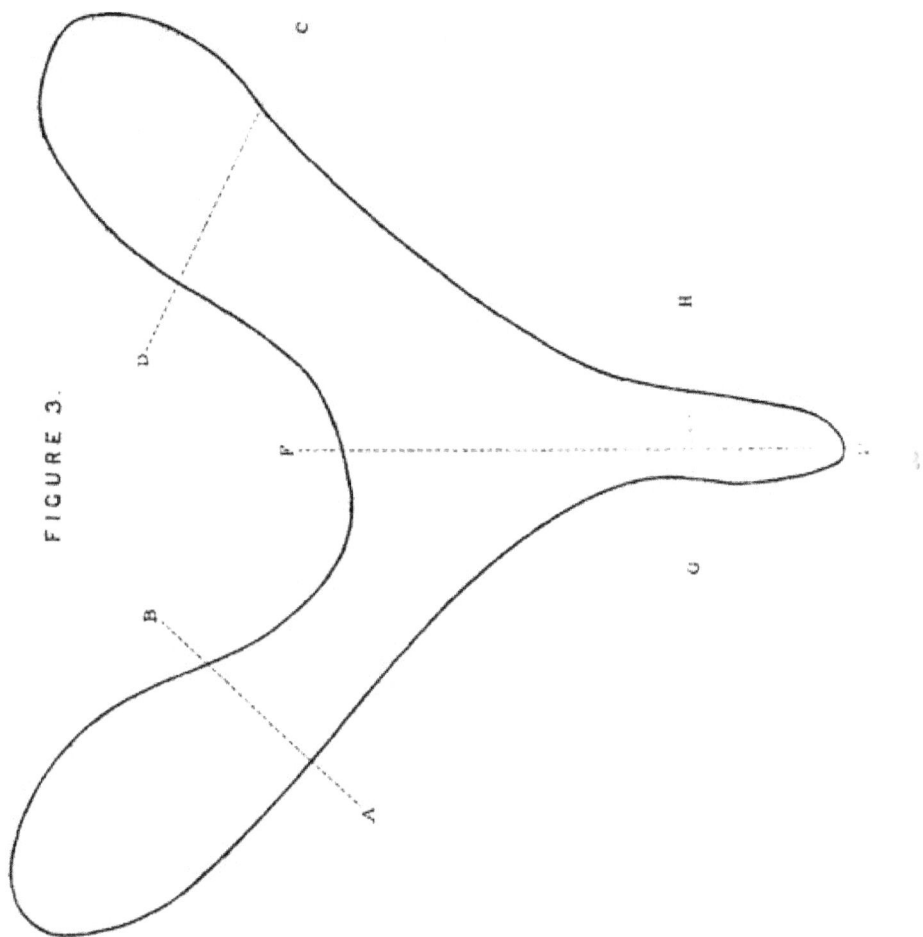

FIGURE 3.

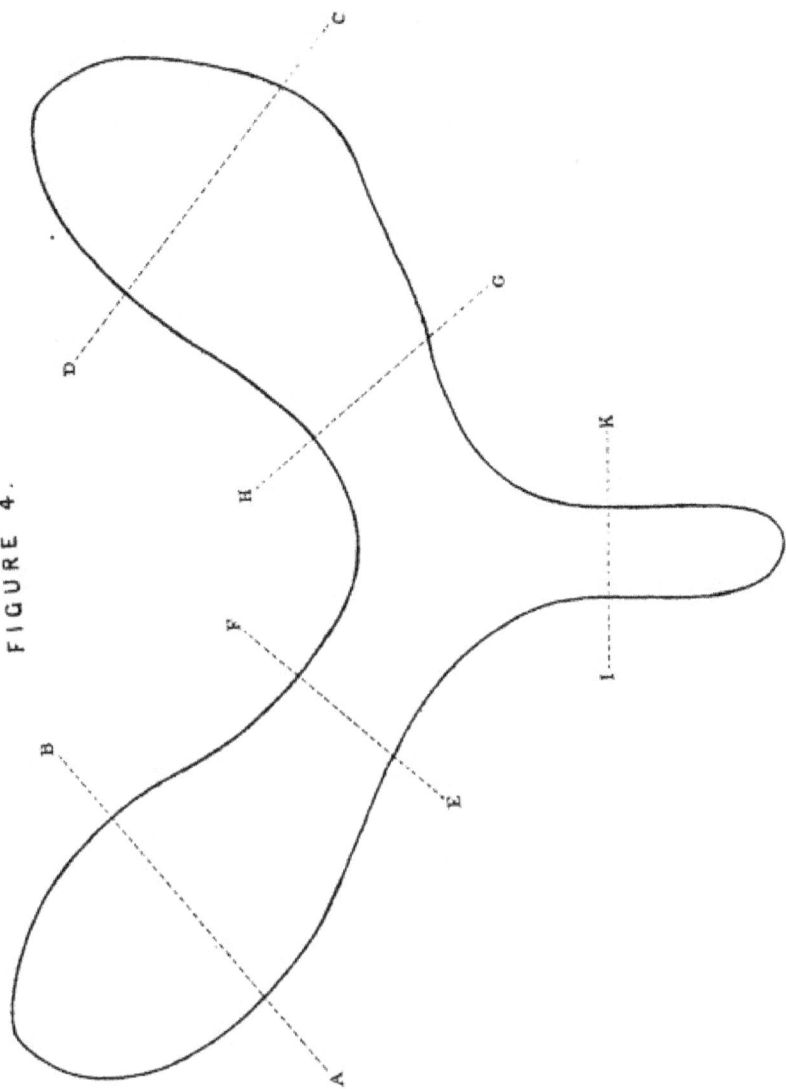

FIGURE 4.

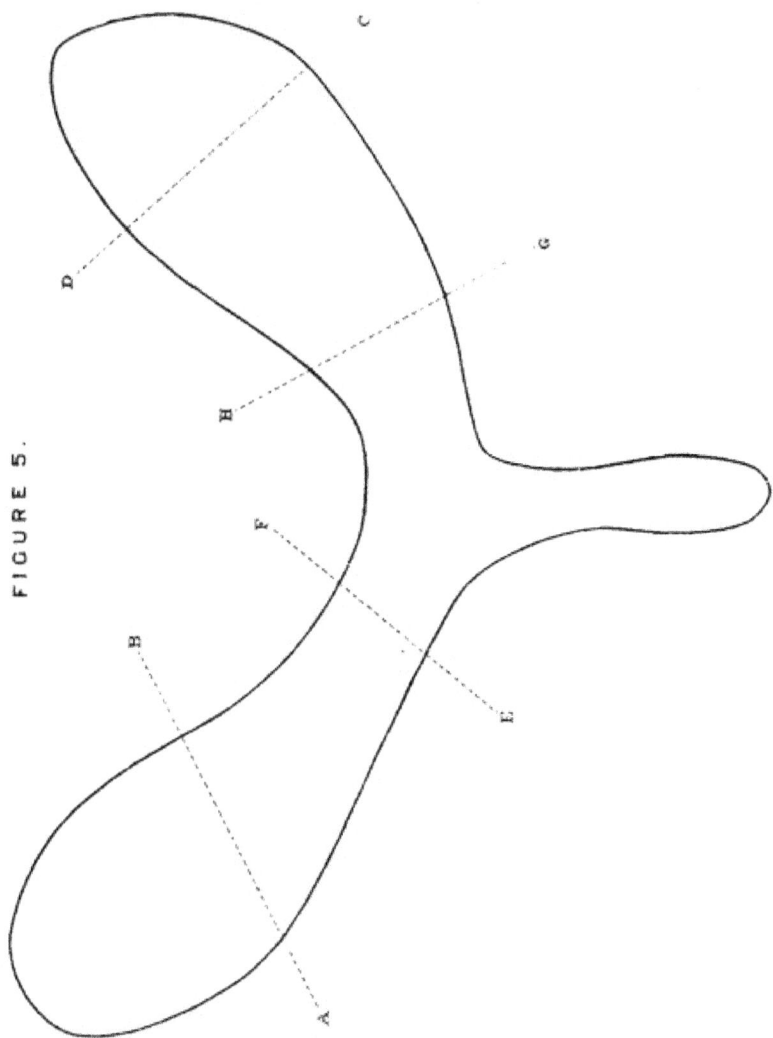

FIGURE 5.

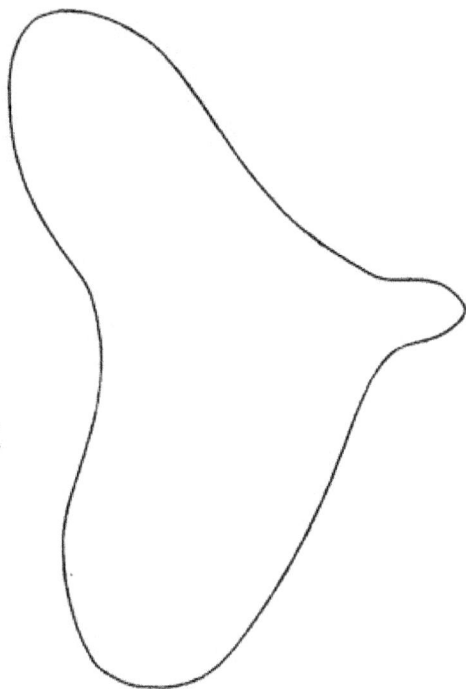

FIGURE 6.

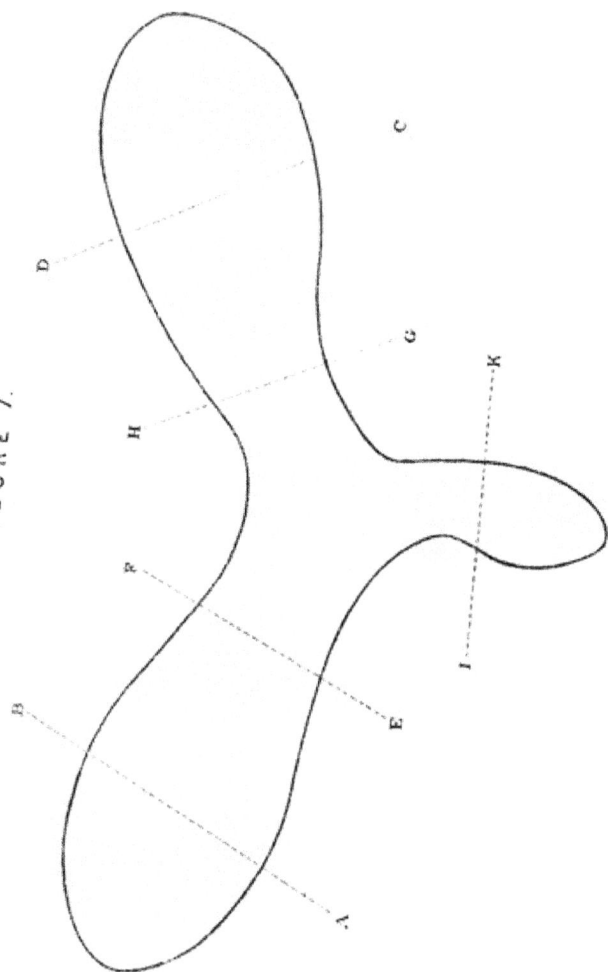

FIGURE 7.

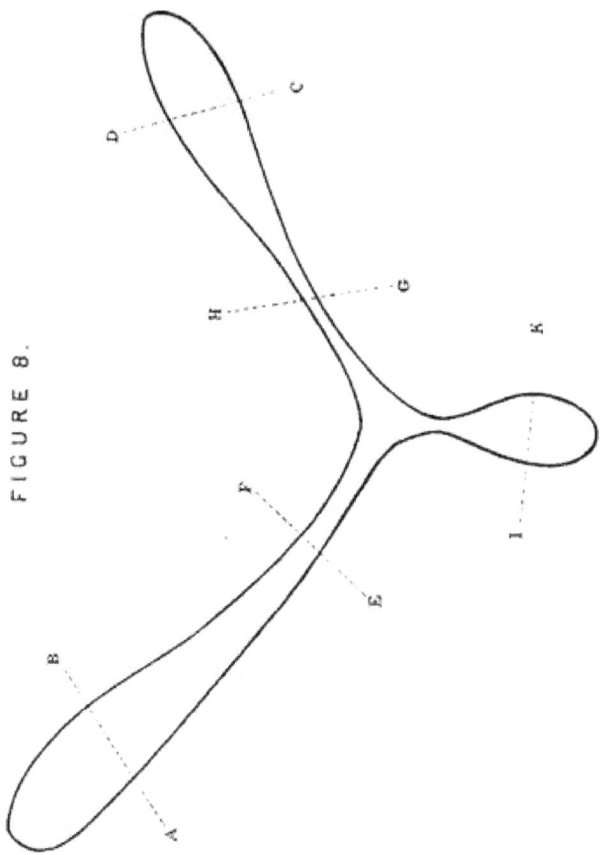

FIGURE 8.

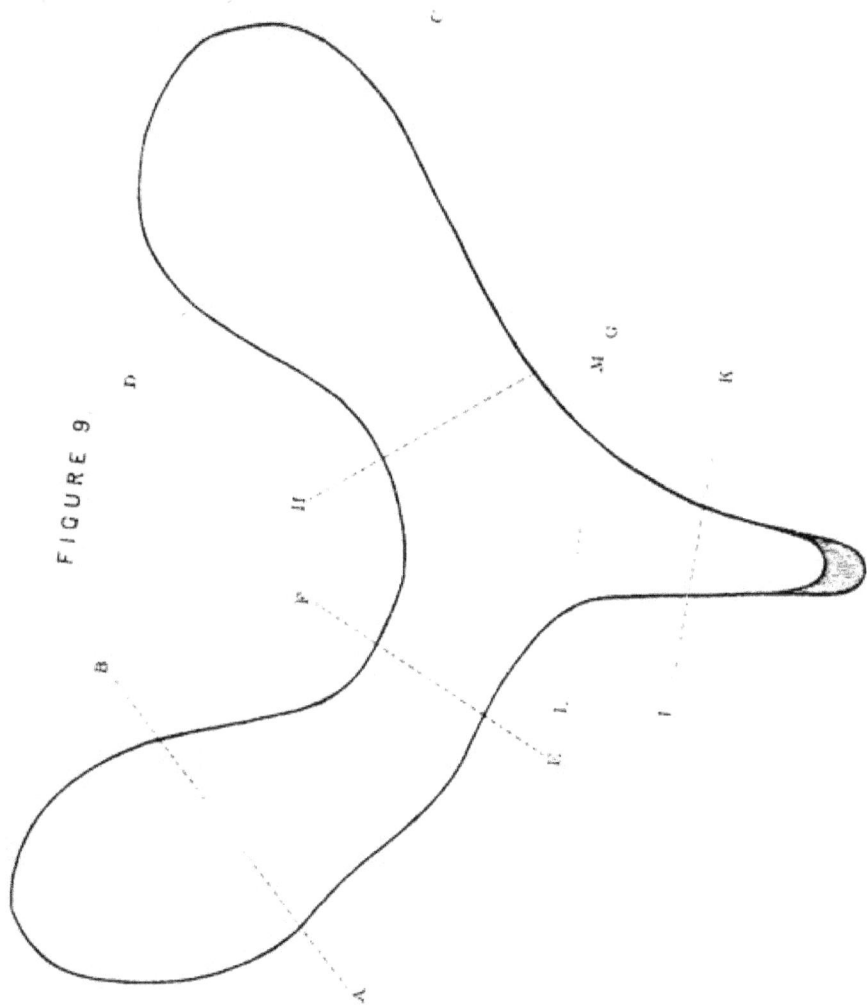

FIGURE 9

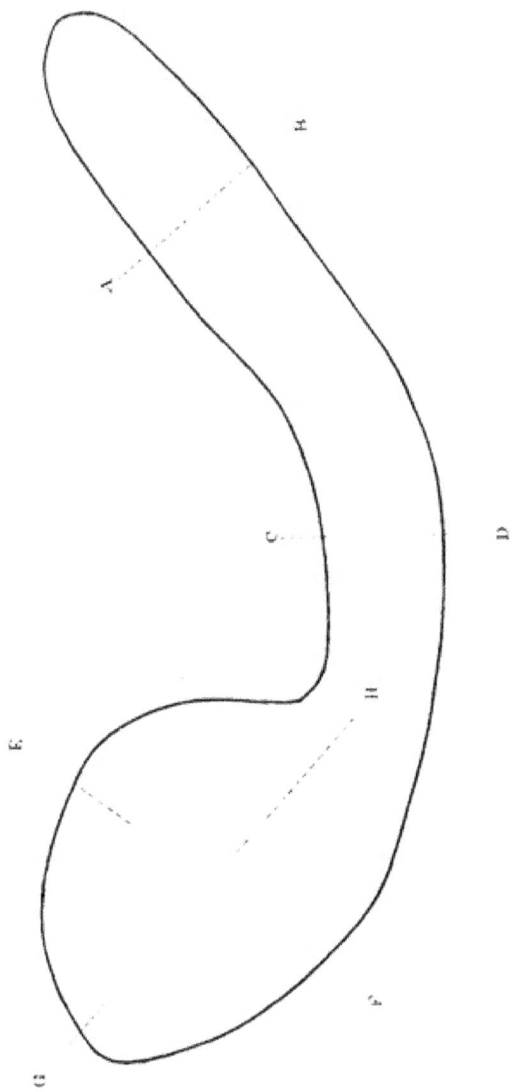

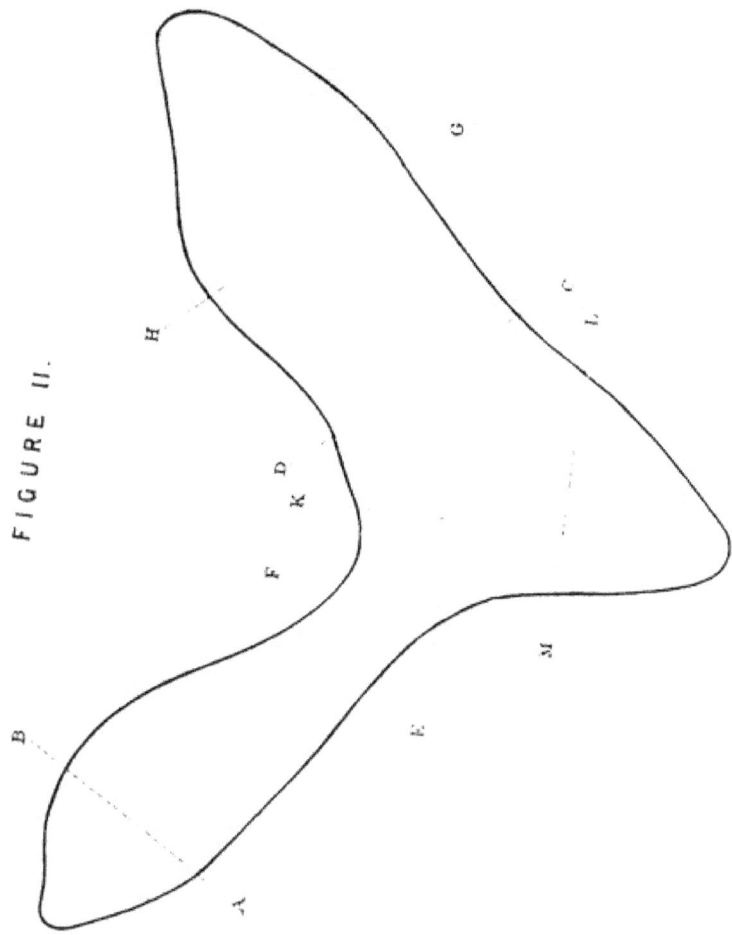

FIGURE II.

PART I.

INTRODUCTORY REMARKS.

The Cæsarean section is not an operation of recent date; its performance is recorded before obstetric medicine and surgery were scientifically accepted (vide *Edinburgh Medical and Surgical Journal*, vol. xxv.), and has since been generally recognized in most of the modern systems of obstetricy. Although this is the fact, yet it has not received the unanimous approval of the members of our profession. From a very early date it has had its advocates and its opponents. To my knowledge, there has been no subject connected with medicine which has created more bitterness of feeling and animosity in the minds of those who may be classed as Cæsareanists and anti-Cæsareanists.

In no city or town in these empires have these repugnant and unprofessional feelings existed to a greater extent than in Manchester. The important but rancorous controversy which took place here between Dr. Hull and Mr. Simmons brought the greater part of the medical profession to entertain more clear and definite opinions. The writings of Dr. Hull, apart from their controversial character, contain most valuable and practical observations.

When I received the honour of appointment to deliver the first obstetric address before the Provincial (now named British) Medical Association, at the next meeting which was to take place at Manchester, I selected a practical and at that time a debateable subject; and, even at the present time, the opinions of medical men are unsettled and discordant upon it. My opinions upon some parts of the subject, no doubt, differed from the great majority of those who honoured me by patiently listening to its delivery; yet I did not hesitate freely, and I

hope conscientiously, to express them.   At the present time
I do not shrink from the responsibility of again bringing more
fully before the entire profession my views, which had only
been partially known for several years before the delivery of
the address.   I have the fullest confidence that the doctrines
promulgated will receive the unprejudiced judgment of the
profession.

I was induced to select for my subject the Cæsarean section,
and those other means which have been recommended to
supersede its performance, partly because these subjects have
been, as already stated, warmly discussed in this city; and
partly because the greatest number of cases (I speak relatively)
in which the Cæsarean operation has been performed in Great
Britain and Ireland have occurred in this city and in the
neighbouring districts.   The analytical tables contain a report
of seventy-seven cases.   Of this number, fifty-five have happened in England—of which, twenty-five have occurred in
Lancashire; fifteen cases have occurred in Scotland, and seven
cases have taken place in Ireland.   It is a remarkable fact
that there stands no case recorded from Wales.

The following observations are entirely confined to British
and Irish cases.   I have purposely avoided admitting foreign
cases into the tables, or of making remarks upon, or of drawing any deductions from them; although I am quite aware the
maternal mortality might be shown to be considerably less by
their admission for computation, than it appears by only taking
the results of British and Irish practice.

In the following pages all the questions have been faithfully and conscientiously discussed; and all the opinions which
are given are, as far as possible, based on facts.   My object
is to endeavour to place the Cæsarean section, and some other
obstetric operations, on such medical, social, and moral
grounds, as to be approved by both the profession and society
at large.   The doctrines which I have inculcated in the following pages are only desired to be received in the spirit in
which they have been written; and I desire them to be taken
in no other way than as they are worthy of acceptance or
rejection.

The tables which were brought before the Association contained records of many points which have been now omitted

in order to reduce them. They contained an account of the number of previous labours, and the mode of delivery; the state, &c., of the os uteri; the location of the placenta, &c.; the exact line of the incision; and some general remarks on the condition of the patient before, during, and after the operation, &c. I have, however, embodied the deductions to be drawn from the record on most of the subjects above adverted to.

CHAPTER I.

ON THE NECESSITY OF THE CÆSAREAN SECTION AS AN OBSTETRIC
OPERATION.

In an ill-directed controversial ardour, Mr. Simmons, in his remarks addressed to Dr. Hull, declared that the Cæsarean section was universally and inevitably fatal, and proposed a compound operation of symphyscotomy and craniotomy to supersede it. It was not long afterwards before he had an opportunity of putting in practice his highly lauded operation. He was consulted in the notable case of Elizabeth Thompson, whose pelvis (now in my possession) is extremely distorted. An examination of it must have brought conviction to his mind that some other means must be adopted in order to deliver her. It is to be presumed he renounced his proposed operation, as he discarded his patient, who was afterwards brought to the Manchester Lying-in Hospital, and delivered by the Cæsarean section.

Dr. Hull, at this time, endeavoured to settle the disputed question of the necessity of this operation ; and the soundness and justice of his opinions were generally approved of by the profession. If this question had still remained undisturbed, it would have been unnecessary for me to interfere with this part of the subject. Within a few years, however, not only the necessity, but likewise the propriety, of its performance has been denied, and opprobrious epithets employed (unworthy of the talented physician who used them), which, although totally unfounded, not only cast odium on the operation, but also reflect most unjustly on the character of those obstetricians who have conscientiously recommended it and boldly performed it. More recently, it has been declared by an obstetric physician, that the induction of abortion, the induction of premature labour, craniotomy, or these two last operations combined and applied according to the degree of distortion, would render the Cæsarean section altogether unnecessary.

The Cæsarean section is doubtless required whenever the pelvic apertures, or its cavity, are so diminished that a mutilated infant cannot possibly be drawn through. This diminution may be positively produced when the pelvic bones are distorted by mollities ossium, by rickets, or by irregular union of, or by a large deposition of callus on, these bones after fractures; or from exostosis, which may grow upon any portion of the bones.

The pelvis is also sometimes relatively so diminished by different kinds of large tumours which are lodged within its cavity; some of which are loose, whilst others are immovably fixed so as to render this operation necessary.

An analytical statement of the causes which have rendered the performance of the Cæsarean section necessary in these kingdoms, will be found numerically to stand as follows. Of the seventy-seven tabulated cases, forty-three were produced by mollities ossium, of which thirty-two were English, ten Scotch, and one Irish. In fourteen cases the pelvis was distorted from rickets, of which twelve were English and two Scotch.

In one case the distortion was congenital (English), and was of a rickety character; in two cases, one English and one Scotch, the pelvis had been fractured.

In six cases fibrous or other tumours existed in the pelvis; of which three were English, two Scotch, and one Irish. In two cases there was an exostosis growing from the base of the sacrum; of which one was English and one Irish. In two English cases, carcinoma of the os and cervix uteri caused the obstruction. In seven cases the cause is not recorded.

Nearly all the pathological causes enumerated above, which render the Cæsarean section necessary, are progressive; and most of them may proceed to such an extent as nearly to obliterate the apertures of the pelvis, or to block up the cavity.

I have in my possession a distorted pelvis in which the brim is nearly destroyed, there not being a greater space between the descending lumbar vertebræ and the pubes on each side than the tenth part of an inch. The space between the lumbar vertebræ and the rami of the pubes is five-sixteenths

of an inch, and the space between the jutting of the pubes
near the symphysis is three-eighths of an inch.

So, likewise, exostosis, or tumours within the cavity, have
grown so large as to prevent a finger from passing without
great difficulty. These pelvic conditions may exist in a first
pregnancy, or may come on at any time during the child-
bearing period; and a woman who has had several natural
and propitious labours may, in successive cases, have greater
or less impediments existing, which may require different
means for her delivery; or the pelvis may be naturally capacious in one labour, and in her next the bones may be so
distorted, or its cavities may be so filled, as to require the
Cæsarean section.

Then, with such uncertainties as these, it is obvious that both
the patient and practitioner may be completely ignorant of these
organic conditions until pregnancy has either been considerably
advanced, or even completed, and labour commenced.

Surely, the most benighted opponent to the Cæsarean section
cannot be so mentally blind as not to know that young married
women cannot be compelled to submit to vaginal or other
examinations in order that it may be ascertained whether there
is sufficient pelvic capacity for a full-grown infant to pass
through. But, supposing the practitioner to be acquainted
with the state of the pelvis, the means recommended to supersede the performance of the Cæsarean section are quite inadequate to prevent its necessity in those higher states of distortion, &c., which have existed in most, if not all, of the cases which
are tabulated. Ample testimony exists of the truth of the
above remark in some of the cases contained in the tables.

Dr. Hull relates several cases, and within my own knowledge
others have occurred, in which it was quite impossible to deliver
the women after either embryotomy or craniotomy had been
performed. Then, under these circumstances, what measures
must be adopted for the delivery of the woman? Must she
die undelivered with a mangled infant still in the womb?
This event has been most unwarrantably allowed to happen.
Again, ought the Cæsarean section to be performed to extract
a mutilated infant? This practice has been pursued. These
are weighty reasons why the Cæsarean operation should be
considered as one, at least, of necessity. There are, however,

other grounds to be spoken of, which further establish this proposition. No doubt every obstetrician will admit that it is absolutely necessary for the os uteri to be accessible when he intends to induce abortion; more especially, if this operation is to be performed by the aid of instruments. And when, at later periods of pregnancy, craniotomy is contemplated, it would doubtless be considered a *sine quâ non* that both the os uteri, the degree of its dilatation, and also that the presentation of the infant, should be ascertained before this destructive operation is performed. These important desiderata do not, however, always exist in cases in which the pelves are highly distorted.

In twenty-one of the tabulated cases the os uteri could not be felt; in twenty-one cases there is no account given, from which it is fair to conclude that it could not be touched— making together forty-two cases. In thirty-five cases the os uteri was discovered with more or less difficulty. In sixteen cases no part of the infant could be reached. In forty-one cases we have no account; which omission affords negative or presumptive evidence that the presentation could not be ascertained, which together make fifty-seven cases. In twenty-one cases the following presentations are recorded: in twelve the head, in three the hand, in two a hand and a foot, in one a foot, in one a hip, and in two the arms.

The foregoing remarks, and the above authenticated facts, are, I hope, amply sufficient to establish the proposition of the necessity of the Cæsarean section as a recognized obstetric operation, although the subsequent observations do not relate to the necessity of the operation, yet I deem them so practically important that I have ventured to place them in this chapter.

In some cases of rupture of the uterus, the infant might be removed with more safety to the mother by an abdominal section, than by dragging it away either by the feet or by the crotchet. Trask's extensive statistics are very favourable to its adoption in some cases of this accident.

When a woman who is nearly at the full period of pregnancy dies, or is killed by accident, the obstetrician is, morally, socially, and professionally, bound to propose *post mortem* hysterotomy. Justice to the incarcerated (most likely living) infant demands an immediate decision, as too long delay would be hazardous to its life. It is, however, a well-known fact that the infant survives the death of its mother much longer than is usually supposed. In this empire medical men are quite at liberty to exercise a free and conscientious judgment; they are not trammelled by theological dogmas, as they are in France and other countries.

CHAPTER II

THE statistics of the results of the Cæsarean section, especially as concerns the mothers, are highly unfavourable. The general account stands as follows of the seventy-seven women whose cases are tabulated. Sixty-six, or 85·71 per cent., died; eleven, or 14·28 per cent., were saved.

The number of successful cases here mentioned is greater than is usually allowed to have taken place; and, therefore, this statement requires further explanation. They are registered as follows. Nos. 1, 12, 35, 36, 37, 49, 53, 57, 67, 68, and 71; of these, Nos. 1, 12, 36, 49, 53, 67, 68, and 71, perfectly and permanently recovered. Case No. 35—She also recovered; the wounds being nearly healed. She lived several weeks; but afterwards she died from epilepsy, to which malady she had been previously subject. Case No. 37—The woman recovered; and afterwards died from disease of the hip-joint.

There is, however, another case included in the deaths which ought, in my opinion, to be in some measure considered as one of recovery. She lived seven days; and so long as she was rationally treated she went on favourably. But after the treatment had been injudiciously changed she gradually grew worse and died.

The special statistics, or the results, of the cases in which I have been concerned are as follows. Of six women, four died, or 66·66 per cent.; and two were saved, or 33·33 per cent.

From the seventy-seven women, seventy-eight infants were extracted (one being a case of twins), of which forty-six, or 58·97 per cent., were saved; and thirty-two, or 41·02 per cent., were dead. Nearly all these infants were dead before the operation, which might have been saved if it had been earlier performed.

The special statistics, or the number of deaths, in my prac-

tice stand thus.   Of six infants extracted, three, or 50 per
cent., were saved ; and three, or 50 per cent., were dead.   Two
of this number were dead before the operation, one of which
was putrid ; the death of the other was doubtless chargeable to
the operation, and was caused by a spasmodic seizure of its
neck by the uterus during its extraction.

The risk to infants in Cæsarean births is not much greater
than that which is contingent on natural labours, provided
correct principles of practice are adopted.

If I dare venture to give an ideal comparative estimate, I
should say, if it is supposed 1 per cent. be the mortality of
natural labour, that consequent on the Cæsarean section may
be stated as scarcely 1½ per cent.

CHAPTER III

HAVING, in the preceding chapter, placed before my readers a full and trustworthy statistical account of the results of the Cæsarean section in the cases in which this operation has been performed in Great Britain and Ireland, I shall next endeavour to prove what the causes are which have occasioned such a fearful fatality of the mothers, and how far they unavoidably belong to the operation. I shall then speak of the infantile deaths and their causes.

To satisfactorily and faithfully accomplish this investigation, the mind ought to be free from all partiality in favour of the Cæsarean section and from all prejudice against it. The deductions on which we seek to establish practical principles ought, as far as possible, to be drawn from well-established facts. However true this rule in general is, there is more or less difficulty in strictly observing it on the subject now under our consideration. Most of the cases, in my humble judgment, have been related more for the object of swelling the already fatal list, than for the purpose of pointing out the mischief which existed previously to the operation, and the real causes of death.

I. *On the Causes of Maternal Mortality.*—The constitutional state of most of the women who underwent this operation was very unfavourable for its performance. Forty-five of them laboured under progressive and incurable disease; many of them were bedridden, and were also unable to discharge their social duties. Many others wanted that perfect or conservative constitutional power to enable them to bear without danger so important an operation.

In most of these cases the practitioner was ignorant of their nature until his assistance was required during labour, and therefore he could not adopt preparatory measures. In all capital operations the risk is greatly enhanced if such means have been neglected. The blood must be depraved in such subjects, and consequently the secretions and excretions must be unhealthy; hence the necessity of taking such steps as tend to correct organic or functional derangement. Constipation is nearly an invariable attendant on ordinary pregnancy; and, in many cases, fæcal accumulations to a great amount occur. But when distortion of the pelvis exists, this is much more likely to happen, in consequence of the mechanical impediment offered by the great projection of the promontory of the sacrum and lower lumbar vertebræ to the downward passage of the fæces. The cervical and oral portions of the uterus, which are thrown backwards against this osseous mass, tend to compress the intervening gut. The same effects, to a greater or less degree, are produced when large tumours exist in the pelvis. The numerous evils which arise from neglected bowels are not only experienced during pregnancy, but also during the puerperal state. Such are peritoneal inflammation, puerperal irritation and exhaustion, &c. If, then, such serious diseases occur during the puerperal state after ordinary labours, from causes which are remediable, is it not very probable that the same mischief might happen after Cæsarean cases, in which these causes do exist in a still higher degree?

Labour, if unduly protracted, is nearly always attended and followed by a considerable number of very serious evils.

These mischievous effects vary considerably according to the duration of the labour—to the nature of the cause and the degree of the mechanical impediment which obstructs the passage of the child through the pelvis. And, therefore, it is obvious, different measures must be adopted according to the relative degree of obstruction. We ought, however, always to consider a lengthened duration of labour, from whatever cause it arises, as more or less unfavourable to both the mother and her infant. In all such cases we should be extremely watchful, and timely adopt those measures relatively required for the delivery of the woman before any injury is inflicted on, or irreparable mischief is done to, the pelvic tissues or organs. It must be understood, that all the dangers of protraction increase after the rupture of the membranes and the discharge of the liquor amnii. It is also a well established fact, that the dangers both to the mother and to the infant increase in a ratio proportionate to the duration of labour. I soon learned, from my hospital practice, that the rules laid down by systematic writers on midwifery, on the treatment of protracted labour, were most mischievous.

To the students of my class, I invariably and urgently inculcated the necessity of an early performance of all obstetric operations, either manual or instrumental, as being of the highest importance, and as especially tending to save the lives of both mother and infant when those instruments were used which are compatible with its life.

In the year 1843 I delivered a short course of lectures to many members of our profession, in which I urged the propriety of an early performance of all obstetric operations, especially of the Cæsarean section, and pointed out the progressive dangers of protraction. At this time I had no tables to guide my opinion, with the exception of those of Dr. Breen in his observations on the management of tedious labour (*Edin. Med. and Surg. Journal*, vol. xv. p. 161). These tables clearly show that dangers increase with the duration of labour. Since this period, Professor Simpson has most satisfactorily proved this fact. It may not, perhaps, be considered irrelevant briefly to mention the effects of labour when unduly and unwarrantably prolonged, in order that a comparison may be made between them and those which have been found existing after the Cæsarean section, and which have been most unjustly attributed to this operation.

Sometimes febrile excitement occurs, accompanied with a quick pulse, hot skin, great thirst, and furred tongue. If means of relief be not afforded, more alarming symptoms soon follow. The tongue becomes covered with sordes ; the pulse becomes more feeble ; and sinking and exhaustion take place, followed by death. Apoplexy, or hæmorrhage from the lungs, may occur in women predisposed to these diseases ; or the large vessels of the heart may suffer. Atony of the uterus happens, giving rise to flooding. Active or sudden rupture of the uterus frequently happens. There often takes place a destruction of tissue in the cervix uteri, from the contusion which this part sustains by the forcible pressure of the child's head against the pelvic bones. The os uteri is sometimes separated from the cervix. In other cases gangrene of the cervix and os has taken place. Inflammation of the cellular tissue of the pelvis occurs, with its consequent infiltration, suppuration, and abscess. At other times the textures of the different pelvic organs are destroyed, and sloughing takes place, which makes intercommunications between the vagina and the rectum, or between the vagina and the bladder, constituting the recto-vaginal or vesico-vaginal fistula, with a train of evils which make the life of the woman most miserable.

The nervous system may be considerably influenced by the Cæsarean section, as it is by most, if not all, other capital operations, the effect of which is termed " shock." This has been asserted to be a frequent, an unavoidable, and an uncontrollable cause of the woman's death. If an abstract view only be taken of the condition of the patient after the operation, then this statement would in some measure appear to be true. But a careful consideration of all the preceding contingent circumstances which existed in each of the recorded cases, and more especially of those which have occurred to myself, leads me to a different conclusion. All the patients in whose cases I have been concerned bore the operation with great fortitude and moral courage, and some of them expressed themselves as having endured less pain than they had felt from one of the labour-throes. There was not any manifestation of shock produced by the operation which did not exist before its performance. If women who had not been endangered by previous disease, or who had not suffered from the effects of protracted labour, died suddenly, or in a few hours after this operation, without any rally, then it would be reasonable—nay, quite right—to attribute their deaths to the shock occasioned by it. But the fact is otherwise, as nearly all those patients registered in the tables laboured under an incurable disease, and had been a considerable time in labour. I here insert the durations of the labours in sixteen of the tabulated cases, in which sinking, exhaustion, or the effects of shock are stated as the real cause of death. In one, it was twelve days ; in one, it was ten days ; in one, it was seven days ; in one, it was six days ; in one, it was four to five days (in this case turning had been unsuccessfully attempted, and afterwards craniotomy ineffectually performed, during which operation the vagina was lacerated); in one, it was a hundred and two hours; in one case, it was three days and a half; in three cases, it was three days (one of these women died from disease of the lungs); in one, it was sixty to seventy hours; in one, it was sixty hours; in one, it was thirty-six to forty hours; in one, it was thirty-five hours; in one, it was thirty-four hours; in one, it was twenty hours. One was only twelve hours in labour. She was greatly reduced in vital power by unavoidable hæmorrhage (placenta prævia); she had also bronchitis and epileptic convulsions both before and after the operation.

These cases require no further comment than to say they afford sufficient evidence of the real cause of death, which truly cannot be attributed to the operation.

Hæmorrhage with shock is stated to have been the cause of death in some of the tabulated cases. The duration of labour in these women is noted as follows. In one, it was fifty-four hours; in one, fifty-five hours; in one, fifty-six hours; in one, seventy-two hours; in one, thirty hours. In this case, there had been a considerable loss of blood before the operation; but very little was lost afterwards. Embryotomy had been unsuccessfully performed, the uterus ruptured, and the os separated from the cervix uteri. In one case, the labour lasted eighteen hours. There was very little blood lost during the operation; but internal hæmorrhage afterwards took place. There were three pints of blood found.

Hæmorrhage has been alleged to be one of the causes of the fatality of this operation. Dr. Hull, however, in two or three parts of his controversial writings, denies that serious danger occurs from this cause; but a strict analytical inquiry of the tabulated cases proves that this assertion is not correct, but that a greater or less quantity of blood is sometimes lost. In several instances, the discharge was considerable, and perhaps may be said to have been dangerous. The peculiar sources whence blood issues during this operation are from the incised edges of the abdominal and uterine parietes; and, when the placenta is in the way of the incision, it may be cut, and then blood issues from its divided structure. Hæmorrhage sometimes proceeds from the uterine arteries, and from the large sinous openings, and also from the surface of the placenta when it is partially separated; and, when this organ is torn, blood is discharged from its disrupted textures, as happens after ordinary labour. In the seventy-seven cases, it is recorded that in twenty there was no blood lost; in twenty-four, very little was discharged, varying from two to seven ounces in quantity; in five cases, there were seven to ten or twelve ounces; in four cases, there were fourteen to twenty-four ounces discharged. In twelve cases, the extent of loss is not definitely stated; but the following expressions are used, as "very considerable," "profuse," "a gush," "really frightful," "not alarming," "great and welled up." These statements are so vague as to be completely valueless, and cannot enable us to judge whether the patients were really endangered by it. We know too well what varying accounts are given by different persons as to the quantity of blood lost on ordinary occasions, to receive the above terms as evidence of a positively serious loss. It is very probable that the amount of blood lost in most of these cases did not exceed that which is discharged after ordinary labours.

In four of the cases, chloroform was administered; and in one, etherisation was used before and during the operation.

In twelve cases, the placenta was cut upon; in one of which there were twenty to twenty-four ounces of blood lost; in one, fourteen to sixteen ounces; in one, ten to twelve ounces; in two, a considerable quantity was lost; and in seven or eight, the quantity was very trifling.

In two cases, the epigastric artery was divided; but there was little bleeding, and it was readily stopped.

In seven cases, the blood issued from the uterine tissue during the incision.

It has been asserted that these accidents (in Cæsarean cases) depend on causes which are not very much within obstetric control. This statement is, however, very far from true. In the majority, the sources whence the blood flows are, as has already been mentioned, the same as those whence it issues in other cases of flooding.

The complete contraction of the uterus must, if possible, be obtained; and the placenta must be promptly removed. The latter part of this rule can always be easily and effectually carried out; but there is more difficulty to fulfil the former part of it, as the contractility of the organ is considerably impaired. This is a common effect after protracted labour. In most of the Cæsarean cases, the operation was not performed until the power of the uterus was completely worn out; and in many cases its tissue was disorganized. The relative and comparative tolerance of the loss of blood in such cases should be duly considered; and as far as possible this accident should be guarded against. It is not, however, true that "the resources of art can effect but little," or to look upon it as a certain contingency upon the operation.

The vital powers of most of those women who underwent the Cæsarean section were at a very low ebb previously to the commencement of labour, and were further seriously exhausted by its duration.

Inflammation of the peritoneum is considered as a frequent cause of death after the Cæsarean section. This serous membrane is usually more susceptible to morbid disturbance after labour than it is at ordinary times; so that it cannot be wondered at, that this disease is sometimes found in Cæsarean cases, especially if the previous management of the labours, and the real condition of the patients at the time of the operation, are duly considered. The duration of the labours of the women who had peritonitis is as follows. In one, it was six days; in one, three days; in one, sixty-one hours; in one, sixty hours; in one, fifty-three hours; in one, fifty-two hours; in one, forty-eight hours; in one, forty hours; in one, thirty-eight hours (craniotomy was unsuccessfully performed in this case); in one, thirty hours (turning was unsuccessfully attempted, and craniotomy was afterwards ineffectually performed); in one, eighty-two hours (attempts were unsuccessfully made, the membranes being ruptured, to induce premature labour; and afterwards craniotomy was performed without success). Other periods are to be noted—twenty-four, twenty-nine hours, &c.; and in one the labour lasted only thirteen hours, but in this case the uterus was stitched with the glover's suture. In one woman, the duration of her labour was fifty-four hours; she had flooding before the operation. In another woman, whose labour lasted ten days, it is stated that she had peritoneal inflammation, in conjunction with the effect of shock.

In two or three cases, the lower portion of the cervix and the os uteri were gangrenous. The violent and constant pres-

sure of these portions of the uterus betwixt the head of the
infant and the irregular projection of the distorted pelvis, cannot
be unlimitedly continued without producing either laceration
or disorganization, or complete destruction, of their tissue.

It is alleged that the abdominal and uterine wounds have
been found in very different states, most of which are said to
have shown a feeble reparative power and a perverted action.

When the vital powers are good in cases of an operation, we
have conservative and restorative action immediately set up;
but, if they are impaired by the existence of positive disease,
or by protracted labour, unhealthy action takes place; and,
instead of healthy surfaces, flabby and œdematous edges of the
wound are seen; and, instead of an adhesive effusion, there is
a dirty sanious discharge, and the uterine wound is said to be
found generally gaping. Although remarks on the bad state
of the wounds are made to depreciate the value of the Cæsarean
section, yet it cannot be a matter of surprise that they should
sometimes present such very unhealthy aspects, if the previous
constitutional and local condition of nearly all the women
upon whom it has been performed be justly considered.

Another cause of danger is supposed to be the reduction
which goes on in the puerperal uterus, so as to regain its
pristine size. This change is considered antagonistic to the
reparative action necessary to heal the wound made in the
Cæsarean section.

The entire vascular system of the puerperal uterus is very
considerably altered; and the supply of blood to it is consequently very much lessened from that which existed during
its gravid state. This change, together with a process of
absorption which is now set up, at least partly causes its
diminution of size. But this organ may be, as it is stated,
reduced in bulk by a general degradation of its tissue; of
which the abundant presence of fat-globules in the lochial
discharge, and in the *débris* which covers the interior of the
organ, is ample evidence. These are natural changes, and not
pathological; and, if there be a reduction of bulk in the
uterus, there is simultaneously a relative diminution in the
size of the wound. It has yet to be proved whether the
alteration which the puerperal uterus naturally undergoes will
in any way interfere with the process of reparation.

The want of union of the edges of the uterine wound by adhesion or by granulation is traceable to causes which have already been frequently mentioned, rather than to the natural organic changes above stated.

Tetanus has been considered a cause of maternal death after the Cæsarean operation; but I am not aware that this disease is so recorded in any of the tabulated cases. Professor Dubois told me that a patient of his had an attack of this disease in two or three weeks after the operation, and it proved fatal. This disease, it is said, occurs after other obstetric operations, and it sometimes occurs after abortion.

The bursting open of the wound becomes a cause of danger, by allowing the protrusion of the intestines. The attenuated state of the abdominal parietes, which sometimes exists in extreme cases of mollities ossium, occasions this accident. It happened in three cases, in which no attempts were made either to replace the protruded bowels, or to approximate the retracted integuments. This, however, was a great omission. As an example of the necessity and propriety of returning the intestines, if they unfortunately escape through the wound, I refer to the case (one of recovery) of Mrs. Sankey. (*See* Case.)

Frequent examinations *per vaginam* are often productive of very serious mischief. Inflammation, followed by suppuration and sloughing, are not unusual results. Two or three examples will be found reported, in which great tumefaction of the external genitals and an inflamed state of the vagina existed. These results are alone attributable to frequent and unwarrantable interference.

Long and ineffectual attempts to deliver by the perforator and crotchet are highly dangerous. Contusion, lacerations, inflammation, infiltration, suppuration, and sloughing, are consequences which are not unusually to be found in cases in which violent efforts have been made to drag a mangled infant through a contracted pelvis.

II. *On the Causes of Infantile Mortality.*—The fœtus *in utero* sometimes dies from diseases which occur in its own system, and also from morbid changes in the structure of the placenta, which interrupts the supply of blood from this organ. These causes are not, however, confined in their operation to any particular class of cases. The duration of labour exercises very great influence upon the infant. If the membranes remain entire, and the liquor amnii undischarged, it will endure the continuance and violence of the labour-pains for a considerable length of time without injury. But after this event has happened, there is much more risk of mischief; and the danger increases in a ratio proportioned to the length of time the labour is protracted. The deaths of the infants which have occurred in Cæsarean cases are generally to be attributed to the long continued and violent pressure which they have endured during labour.

There is, however, another cause of infantile death which more especially belongs to the Cæsarean section. I mean the spasmodic seizure of the neck or body of the infant during its extraction through the incised opening of the uterus. In general, there is no difficulty experienced in these cases in withdrawing the infant from the uterus; but sometimes some portion of its body becomes so firmly grasped by the uterus in its passage through the incised opening, that great difficulty is experienced in extracting it. There is, however, more danger to the infant when the neck is seized by the uterine grasp, than when it is held by any other part of its body. In such cases the body of the infant has been most easily brought along until the shoulder had passed, when the neck is instantaneously seized, and so firmly held, as to require long and continued efforts to be made in order to extricate the head.

The fact that the uterus in natural labour is energetically roused to expel the placenta which has been separated, first led me to attribute the seizure of the neck of the child during the Cæsarean section to the partial or complete detachment of the placenta. It has lately been doubted whether this theory will suffice to explain it, as "numerous instances are recorded in which the placenta either protruded through the incision, or was found lying loose in the uterine cavity, and in which no inordinate contraction ensued." I am, and was from the first, aware of the truth of this assertion, from an accurate analysis and tabulation of all the published cases; but, notwithstanding the apparent force of this objection, my opinion remains the same.

There are seven cases (two within my own knowledge) in which this event happened; and, as far as I can ascertain, the placenta has been partially or entirely detached in all these cases, or at least presumptively so.

How far the violent uterine action may resemble the spasm of hour-glass contraction, I cannot determine : yet analogy would lead me to think it did.

There is only recorded one (another case) in which the infant was grasped round the abdomen above the hips; the head, shoulders, and trunk, having been first drawn forth. The child was previously dead; and the only effect recognized was the squeezing out the meconium into the uterine cavity. Turning had been unsuccessfully attempted; and during this operation, it was found that there was hour-glass contraction. Does not the occurrence of hour-glass contraction before, and seizure of the child's hips after, the operation, favour the above mentioned opinion ?

CHAPTER IV.

THE process adopted by Nature in some of those cases of labour in which she cannot overcome the obstacle which prevents the passage of the infant through the pelvis, is, in the first instance, the yielding of the uterine tissue, thereby making an opening for its escape. Afterwards, if the constitutional powers prove equal to the entire process, an incasement of the fœtus is effected by the effusion of lymph. And, after a time, a pointing in some part of the abdominal parietes shows itself, which is soon followed by ulceration, and part after part of the infant passes through the opening, until the whole contents of the cyst are discharged. This is a very slow and hazardous process.

In adopting the Cæsarean section, we in some measure imitate Nature in her attempt to remove the infant, although by a much more safe and an expeditious plan.

This operation ought not to be made one of display. There should only be a very few persons present ; and the greatest quietude should be afforded to the patient. Every cause likely in any way to create unpleasant emotional feeling should be most carefully avoided. These rules were strictly observed in the two successful cases in which I was engaged. It is of the first importance, when possible, to adopt all such measures as will prepare the patient to undergo this operation, by improving the general health.

The bowels should be emptied by a large quantity of warm water thrown into the rectum and colon, by an enema-apparatus with a long flexible tube (like the one used to enter the stomach), so that its extremity can reach beyond the great projection of the sacrum.

The bladder must also be emptied by a catheter, equal in length to that used for the male. This organ is forced downwards and forwards, and lies under the deflected uterus, whereby its cervix is lengthened and compressed upon the pubes. This

altered position of the bladder is particularly to be observed
during the latter month of pregnancy, in cases of pelvic distortion from mollities ossium. Frequent examinations *per vaginam*
have been already shown to be extremely injurious; so that
this practice should not be allowed. In an exploration made
to ascertain the measurement of a distorted pelvis, the obstetrician is compelled to pass his hand completely, and as far as
possible, into the vagina. Anxious to ascertain the state of
the os uteri, the presentation of the infant, and the exact
available space in the pelvis, he prolongs the operation, and
often repeats it. And when consultations are numerous (as is
too common) in these cases, serious mischief is inflicted on the
pelvic organs and tissue. By one effectual examination, every
necessary information may be obtained. The interest of the
patient is best secured by having only a limited number (say two
persons) in consultation.

The operation should be performed on the bed; so that the
patient may be kept as quiet as possible afterwards. In some
of the cases in which the woman was removed to a table, some
untoward circumstance happened.

The temperature of the room should be regulated, and a
genial warmth of the atmosphere maintained.

The uterus projects more or less forwards; and when the
pelvic distortion is caused by mollities ossium, this organ
assumes the retort shape. Its projection is so great that its
normal anterior surface rests upon the thighs of the patient
when she sits, so that the fundus necessarily stands most foremost. Before the incision is made, it is of the utmost consequence to raise the deflected uterus up; or else the fundal
tissue, which abounds with large anastomosing vessels, must
unavoidably be divided. Neglect of this caution has, no doubt,
led to the hæmorrhage which happened in some of the cases.
A division of the structure of the upper part of the fundus of
the uterus must certainly interfere with the regular or efficient
contraction of this organ, and thereby produce a gaping
character of the wound.

When we contemplate the mischievous effects of protracted
labour, and review the unfavourable condition in which most
of the patients have been brought by unwisely procrastinating
the operation, we must at once be convinced how important it

is to perform it early.   The sooner the better it is had recourse to after it is determined upon, either as one of election or one of necessity.

When labour is rendered difficult by great distortion of the pelvis, or by large exostoses, or by large tumours in its cavity, some of those natural organic changes are not to be found which would otherwise guide us, and enable us to judge of its commencement and progress.   To wait, then, in such cases as these for the dilatation of the os uteri is not only a great mistake, but also a very great evil; for, in most of them, this part of the uterus cannot be touched, and, in general, very little dilatation of it does or can take place.

The dangers of delay, on expectant grounds like these, which so frequently happened in the registered cases, ought to guard us against waiting for those indications which cannot possibly be discovered, and induce us to operate early.   As soon as the labour is established, and before or immediately after the membranes are ruptured, is the most favourable time to proceed.   Great advantage accrues from adopting this plan; for the length of the uterine incision would relatively diminish in size, equal to the diminution which takes place by the contraction of the uterus.   Another great advantage arising from this course is, that the danger of protraction would altogether be avoided.   It is a well-known fact that little risk comparatively occurs before the waters are discharged.

Before the incision is made, the location of the placenta should, if possible, be ascertained, in order to avoid its being wounded.   In the seventy-seven cases it is reported as follows. In twenty-nine it was connected to the fore part of the uterus; of this number, in two it was placed towards the fundus; in thirteen it was cut upon.   In ten cases it was adherent on or towards the back part of the uterus.   In thirty-one cases the position of it is not alluded to, and, therefore, it is to be presumed it was posteriorly placed.   In five cases it occupied the fundus; in one case it was found near the left Fallopian tube, and in one case it was attached (placenta prævia) over the os uteri.

This minute inquiry as to the precise fixture of the placenta has not been made merely for the purpose of suggesting rules of caution which ought to be observed before making the incision; but, also, of proving that this organ has not a definite position assigned to it.

It is, then, of the greatest importance to make the incision so as to avoid, if possible, cutting upon the placenta, as considerable danger may accrue from so doing.

The stethoscope will nearly always enable us to avoid these hazards. By it we derive positive information of the infant's life by hearing distinctly the pulsations of its heart, and it affords us negative evidence of the infant's death when no cardiac sounds are perceived through it. The audibility of the "placental *soufflet*" directs us to investigate the quarter from whence the murmurs proceed, and, by attention, we may nearly always assure ourselves in what vicinity of the uterus the placenta is fixed. If this sound is not heard, we have a right to conclude that this organ is not within the reach of the knife if the infant be still alive. If it be dead, no great risk will be incurred if the placenta be divided, as the vascular function of this organ will then, doubtless, have ceased.

The position and direction of the external incision has varied. In fifty-seven cases it has been made longitudinally; in eleven of which number it was made on the right side; in twenty-four cases it was made on the left side, and in twenty-two cases it was made in the centre of the abdomen. In two cases it had a transverse direction on the right side. In one it was made obliquely on the right side. In seventeen cases the situation and direction of the wound is not recorded.

I prefer the wound to be longitudinal, and on the left side.

There are no tissues concerned in the operation which require very slow or nice dissection; therefore, unnecessary tediousness should be especially avoided. If the uterus be slowly incised, the stimulus of the knife instantaneously throws this organ in violent and irregular contraction, which separates the placenta and entails mischief on both the mother and the infant. Every precaution having been taken, we ought to strictly observe the motto, "*Cito et tuto.*" The incision should be made on the body of the uterus, because this portion of the organ is eminently contractile, and ought to extend well towards the fundus, but not into it. It ought not, however, to be carried too far down into the cervix uteri, because

this part possesses dilatable properties which are unfavourable
to a diminution in the size of the wound.

When the uterine incision is completed, there should be no
delay in withdrawing the infant. When it lies in its usual
natural position, with the head over the brim of the pelvis,
then the obstetrician should seize its legs with his right hand
and pass his left cautiously and quickly down so as to embrace
the face on one side, or the hind part of its head. By this
mode a double power could be effectually exerted : one of
traction by the legs, the other by raising the head upwards.

If the breech offer at the incised uterine opening, the practitioner should seize it with his right hand and withdraw it,
and at the same time use his left hand as above mentioned.
If the head lay in proximity with the incision, then it ought
first to be brought forth, and, at the same time, he should pass
one hand cautiously forward along its body so as fairly to
embrace the breech, and act with both his hands as recommended above. These precautionary rules are suggested to
prevent the grasping seizure of the neck or the hips of the
infant, as the case may be, during its removal. (*Vide* remarks
already made.) One or two writers have urged that the head
of the infant should be always first extracted, on the grounds
of being safer for it ; but a conditional practice, according to
its position in the uterus, is by far the best.

The head is most generally situated in the lower segment
of the womb, and, therefore, at some distance from the centre
of the incision. In order to bring it fairly to the opening, it
would produce a great strain on, if not laceration of, the contracted uterine tissue, and create nearly a doubling of the
child upon itself before it could be extracted. And as expe-
dition is required, it would be found that the bulk of the head
was not very readily grasped with sufficient firmness so as to
ensure its speedy withdrawal. Time would be lost, and impediments added. The placenta, with the membranes, should be
also quickly extracted.

Protrusion of the intestines is very apt to occur during the
operation ; this becomes very troublesome to the operator and
distressing to the patient, and a considerable time is consumed
in order to replace them. This accident not only predisposes
to remote mischief, but it immediately tends to depress the
vital powers of the woman. She feels faint and has a sense

of sinking. Every care should, therefore, be taken by the assistants to repress and retain these viscera under the integuments by an extended application of both hands on each side of the incision. It is of the utmost importance that the edges of the external wound should be effectually secured. Sutures or pins ought to be inserted at very short distances, and a considerable extent of the parietes (not embracing the peritoneum) should be included, especially in those cases in which the integuments are much attenuated.

The after-management of the patient must be conducted on recognized medical and surgical principles. Much mischief has been done by active treatment; and it should never be forgotten that, even if it be thought desirable to pursue this plan, it should always be relative to the state of the woman. A negative treatment has been found by me most advantageous. Opium, in full doses if required, should be given.

It is now a general practice to administer chloroform before and during the performance of important operations. If cautiously used, the data already accumulated justify the inference that it is of great advantage to the surgeon, by inducing a state of resistless quietude of the patient. The severity of the pain inflicted by the knife is considerably lessened, and the shock to the nervous system is thereby diminished. In the majority of surgical operations there are no other contingent circumstances relative to the administration of this drug which require the attention of the operator, except the necessity of his having first ascertained whether there exist any organic disease of the heart or large vessels which would be dangerously influenced by it; but it is otherwise when it is proposed to use it in a Cæsarean case.

The incision made into the uterus must be at first necessarily large, to enable the obstetrician to extract the infant and the placenta; but, after their removal, the length of the wound is very considerably diminished by the contraction of this organ, which, if not interrupted, is both instantaneous and energetic, thereby effectually preventing any great loss of blood. It is, therefore, very important to inquire whether chloroform interferes with, or altogether suspends, this normal contraction; or whether it induces this action *de novo*, or strengthens it in intensity.

Chloroform has been inhaled in fifteen Cæsarean cases: in

one of which there was hæmorrhage; in two of which there was very little blood lost; one of which cases, it is stated, was benefited by its inhalation; and in three instances the discharge of blood was considerable, two of which proved fatal. One of these cases, however, recovered. In one of the cases the uterus did not contract much; in one the hand was pressed upon it to induce contraction; in four cases, it is stated, that this organ contracted well. Three of these patients were completely unconscious, and one (which I saw) was only partially under the influence of chloroform.

Ether was administered in one case in which there was some bleeding, but not so much as to be considered to be alarming.

Obstetricians differ in opinion as to the positive effects of chloroform on the uterus. Some say uterine action is retarded by it. Others, again, assert. that it does not interfere with it; and there are others who affirm that it promotes and strengthens it. The data which exist on this subject are very meagre and very contradictory; and, therefore, with such discrepancy of opinion, it is impossible to come to a satisfactory conclusion on this subject, especially in reference to its use in Cæsarean cases, in which it is of the highest importance that the normal action of the uterus should not be disturbed. It would, however, be most advantageous to the patient if she could be safely spared the pain inflicted by the operation; although not one of those women in whose cases I was concerned complained of pain during its performance, but, on the contrary, bore it with great moral courage and fortitude; and most of them observed that they suffered less from the incision than the anguish they had endured from one of the unavailing labour-pains.

The disturbed state of the vascular and nervous system in all those women who have undergone this operation must, most assuredly, render them unfit subjects for chloroform; and, therefore, the deductions which may be drawn from the results in these cases in which the women laboured under incurable disease, and were exhausted from protracted labour, ought not to prejudice us altogether against its use.

We find that vomiting occurred in eleven of the sixteen cases in which chloroform had been administered. In two of the cases the abdominal wound was rent open by the violent efforts induced ; and in several of the others disagreeable effects ensued.   If chloroform do really produce vomiting and its injurious effects, there can be no doubt it ought to be discarded, as it is most important to keep the patient free from all causes which have a tendency to disturb the reparative process in the wound.

So long as the Cæsarean operation is considered only one of necessity, and its performance so unwisely and so cruelly delayed, great risk must attend the inhalation of chloroform. But, if it be made an operation of election, so that women who are in a better constitutional state undergo it, and if, likewise, it be timely performed, then it may be found that great advantage may be derived from the use of this drug ; but, nevertheless, before we acknowledge chloroform as a recognized means for this operation, we ought to be fully satisfied what effects it produces on the uterus.

CHAPTER V.

THE other obstetric operations which require to be now considered are reduced into two classes, one of which includes those measures which are to be adopted compatibly with the preservation of the lives of both the mother and the infant. This division embraces the Employment of the Long Forceps, Turning, and the Induction of Premature Labour. The other class of obstetric operations are those by which the life of the infant, or that of the embryo, as the case may respectively be, must be sacrificed—namely, Craniotomy, Embryotomy, and the Induction of Abortion.

I shall make a few remarks on a suggested plan for Dilating the Distorted Pelvis, and I shall also briefly mention Symphyseotomy.

The objects to be attained by these two classes of operations are very differently estimated by different obstetricians. Some consider that those measures by which the lives of both mother and child are preserved are adopted solely for the purpose of lessening the too frequent employment of craniotomy; others consider them as applicable to prevent the performance of the Cæsarean section. But few practitioners would ever think of having recourse to this latter named operation in cases in which either the long forceps, turning, or the induction of premature labour, could be successfully employed. The latter class of operations are performed with the express object and intention, as far as possible, to supersede the Cæsarean section.

CHAPTER VI

I. *On the Employment of the Long Forceps.*—This instrument most justly takes a high position in obstetricy, because its sole employment is for the preservation of life. It is intended, within a certain range of protracted labour, to supersede craniotomy. In the hands of a discreet and judicious practitioner, it is both a safe and a very powerful instrument. Before its introduction into practice, whenever turning could not be performed, the child was doomed to destruction by craniotomy. The employment of the long forceps in this country has been very tardily recognized. When I commenced (1817) my professional career, this instrument had never been used in Manchester; but, having heard the valuable remarks of Dr. Haighton upon its use, I availed myself of the first opportunity of making trial of it. I employed his instrument; but, after repeated trials, I abandoned it, and contrived one of my own, with blades of equal length, but with parallel shanks. This instrument I also found tended, in its embrace and compression of the infant's head, to produce disagreeable effects upon it, which I endeavoured by a second contrivance to obviate. This instrument is so constructed that only a limited degree of compression can be exercised. It has very short handles, and consists of blades of an unequal length : one, the long one, to lie over the face; the other, the short one, to be placed over the occiput. By this arrangement, the head of the infant is placed in the most favourable position within their grasp, and none of the injuries are inflicted upon it which are found when forceps with equal blades and long compressing handles are used. This instrument is employed mainly as a tractor, and very limitedly as a compressor.

To save the life of the child by the use of the long forceps

D

is, doubtless, the object of every obstetrician; for, unless this were his intention, it would be better at once to have recourse to craniotomy. The head of the child cannot bear more than a certain degree of pressure compatible with its life; and, although it is wisely ordained that it can safely bear a greater degree of pressure before than after birth, yet there is a limit even here, beyond which it cannot be carried without the destruction of the infant's life. The head can also bear a greater degree of pressure when the force is applied in one direction, than it can in another. Much greater compressing force can be more safely used when exercised in the bi-parietal than when applied in the occipito-frontal diameter. As the long forceps are usually placed on the head of the infant, so as to embrace it in its long diameter, we ought therefore to consider whether our instrument is so constructed as to permit such an undue degree of pressure as may prove unfavourable to the life of the infant.

The head of the infant, when it is situated at the brim of the pelvis, usually lies with its fronto-occipital diameter corresponding to one of the oblique diameters of the pelvis; the vertex or face being placed towards the right or towards the left acetabulum. But when the antero-posterior diameter of the pelvis is shortened by the sacrum projecting more forwards, the head assumes a more directly transverse position. Now, in this position of the head, it is most desirable to place the two blades of the forceps on the sides of the pelvis, so that one blade lies over the face and the other over the occiput of the infant. In this case the instrument embraces the head in the most unfavourable direction for its safety, if forcible compression be made. But the lateral pelvic position of the blades of this instrument is much safer for the maternal pelvic organs than if, as recommended by some practitioners, they were placed in the conjugate diameter. To add the bulk of the instrument to the already diminished capacity of this part of the pelvis would be unwise. In all our artificial appliances, we ought to endeavour to produce similar changes on the head of the infant which Nature accomplishes if left unaided. The head is lengthened, and its rounded shape changed; whilst its bi-parietal diameter is lessened. The former change we ought to obtain by having the instrument so formed as to allow the head to elongate when traction is used, and by the pressure it receives from the anterior and posterior parts of the pelvis. Notwithstanding the high opinions expressed of the great advantages of compression, I am convinced it is mischievous. This statement is not theoretical, but rests on facts derived from the use of the long forceps both as a strong compressor, and, as now recommended, a tractor with very slight compressing power. In truth, I cannot understand how effective compression can be made, unless the blades of the forceps are applied on the sides of the head, and on the anterior and posterior parts of the pelvis. The tractive power of my instrument is increased by having a handkerchief passed through an opening in the shanks, which is formed by nearly a semicircular curve in each shank near the handles. The handles should be only slightly tied to maintain their position. A pendulum, or side-to-side movement, must be combined with the traction; taking care that the range is regulated by the line of the axis of the pelvis, and that no pressure be thrown upon the maternal pelvic structures.

This instrument is sometimes required to rectify the position of the head of the infant, when its long diameter lies parallel with the antero-posterior diameter of the pelvis; the face lying either towards the pubes or towards the sacrum. In such cases, the blades of the forceps should be introduced along the sides of the pelvis, but should be placed over the parietal bones of the head. (For a further exposition upon these questions, I refer my reader to essays on various subjects connected with midwifery).

There are no statistics published which afford any truthful information either as to the frequency of the application of this instrument or as to the mortality of those women who have been delivered by it. In my own practice I have used this instrument very frequently; and I can most conscientiously assert that I never had a death as a result of its application. In cases in which craniotomy had been performed, in some once, in others several times, under the management of different practitioners, I have delivered the women by this instrument, and saved the children. It is the duty of the obstetrician to keep constantly before his mind the dangers of protraction, and recollect that these increase in a ratio (already stated) proportioned to the length of time the labour is prolonged.

If this instrument is to fully accomplish its capabilities of saving life, it must be used before those dreadful mischiefs are produced by delay. If we calculate to bring a head through a fixed pelvic space, we ought to remember that this space is considerably lessened by the effects of long-continued pressure.

The dogmatic injunctions of present and former authors, that the forceps ought not to be used before the os uteri is fully dilated, or until the woman has been in labour a certain number of hours, or until the ear of the child can be felt, are highly dangerous. They are delusive, and would, if acted upon, altogether prevent the use of the long forceps.

If these alleged conditions are required to exist to guide the practitioner when he ought to have recourse to this instrument, these rules are tantamount to a complete interdiction of its use. In fact, they are too absurd, unfounded, and dangerous, as indications of the propriety of even using the short forceps.

To wait for the dilatation of the os uteri after the rupture of the membranes, is a great mistake; for, in the great majority of cases which require aid by the long forceps, this organic change cannot take place. The obstacle being at the brim of the pelvis, the head of the infant cannot be pressed down upon it; so that, before this change happens, irretrievable mischief may be inflicted by the continued pressure which the pelvic tissues must, under such circumstances, endure. Therefore, as soon as the time has arrived for delivery, we must not hesitate to apply the long forceps, provided the os uteri is so far dilated, and further dilatable, to enable the practitioner safely to introduce the blades.

If the instrument is in the hands of a discreet and judicious obstetrician, no mischief need be dreaded; for the blades of a well-made instrument will rest as safely within the uterus as the hand of the practitioner.

Time ought never to be considered an element of calculation, especially during the second stage of labour, for the use of instruments, except in creating an anxiety to be on the watch, and to take timely steps to deliver the patient.

An early use of the long forceps, when the necessity of the case is established, will prevent those serious constitutional and local mischiefs (already mentioned) which are produced by the long continued pressure of the head of the infant upon the pelvic organs and tissues. Those who are opposed to the use of instruments, and advocate time and patience, attribute to their application those structural and organic lesions which are really the effects of delay.

It has been already stated that the head of the infant can only safely bear a certain amount of compression by the forceps; but it must also be understood that the infant is frequently destroyed by injury inflicted upon the head during protracted labour. Procrastination beyond a certain limit is highly hazardous to its life; and, as its preservation is an important object to attain, the long forceps should be immediately applied (if safe to the mother), if the stethoscope indicates danger.

In convulsions occurring during labour, these instruments may be of essential benefit, if they can be safely used.

In cases of accidental, and also in some cases of partial, attachment of the placenta over the os uteri, this instrument has great advantages over other means, if the os uteri is dilated and dilatable, when the vital powers are very considerably depressed.

In some cases of rupture of the uterus, in which the child does not recede, and if there is sufficient pelvic space, and if all other requisite changes exist for their safe introduction, then there may be a remote chance of saving the infant's life by their use.

In cases of arrest from exhaustion, and even in fatal syncope, this instrument may be usefully employed. In some cases of face-presentations, and other unfavourable positions of the head, the application of the long forceps will be found most advantageous. The want of relative proportion between the infant's head being abnormally large, or too firmly ossified to allow a necessary diminution in its size, in order to pass into and through the pelvis, are causes of protracted labour; and, when the obstacles are so great as to oppose the head passing through the brim of the pelvis, the long forceps will be required to effect the delivery. When the pelvis is relatively too small in its general conformation, from premature defect of development, or when it partakes too much of the male conformation, or in the slight oblique-shaped pelvis of Naegele, or when the pelvic bones have been fractured and there has been irregular union (I have a cast of a pelvis which had been fractured, and which in character now resembles the oblique of Naegele), or when small sized exostoses or loose or fixed tumours occupy the pelvis, the long forceps may be requisite and effectual in the delivery; but, in all these cases, the practice must be determined by the available space which exists in each case.

Those cases in which the value and powers of this instrument are most conspicuous, are those in which the brim of the pelvis is diminished in its antero-posterior diameter, either by rickets or by mollities ossium.

In the slighter pelvic deformities produced by rickets, the brim is often considerably lessened in its conjugate diameter, whilst the cavity and outlet are not very much altered either in size or in shape; so that the obstacle to the descent of the head is chiefly confined to the brim. But, in a pelvis distorted by mollities ossium, the diminution in its capacity is not confined to any one portion, but the brim, cavity, and outlet suffer in a greater or less degree; so that, when delivery by the long forceps is contemplated, the character of the distortion must be well considered.

It must be obvious to every well-informed obstetrician that the head of the infant can be more easily brought through a pelvis distorted by rickets, than through one distorted by mollities ossium; assuming that the antero-posterior measurements are the same in both. Opinions as to the space required to bring the head through the pelvis by the long forceps differ very considerably; and, on reflection, this variation is not to be wondered at. It is at all times difficult to arrive at an arithmetical accuracy by a vaginal examination of the different character of distortion just alluded to. The difference in the size of heads of infants, and likewise the different degrees of ossification, must all tend to influence the result. But, notwithstanding these uncertainties, it is desirable that I should state my opinion as to the smallest space at the brim of the pelvis through which the obstetrician would be warranted in attempting to extract the infant by means of the long forceps. Knowing as I do the great responsibility I incur in making positive assertions on practical points of such importance, yet, as I have by trial proved the truth of my statement, I hope I shall not be charged with temerity. I have the more confidence in giving my opinion on this point, because it is certain that, if this instrument cannot be successfully used, the life of the infant would have to be sacrificed. I would, therefore, rather run the risk of committing a venial error by leaning to the side of mercy, by recommending, in the first place, a cautious trial of the long forceps where there was the least doubt in the mind of the practitioner as to the precise pelvic measurement, and thereby give the benefit of the doubt to the unborn babe. With this feeling, I then say that, where the distance from sacrum to pubes is three inches and a fraction under, and there exists sufficient space in the transverse diameter, an experienced practitioner ought to make a cautious and persevering trial of the long forceps before he has recourse to craniotomy. This opinion is advisedly given, because it is quite impossible to compute either the positive and relative space of the pelvis, or the size, compressibility, or other conditions of the head of the infant, with such mathematical certainty as to warrant any person to destroy life.

II. *On Turning in Cases of Slight Distortion of the Pelvis.*— Turning, in cases of slight distortion of the pelvis, justly ranks as a conservative operation. Turning was formerly had recourse to in all difficult labours in which craniotomy was not performed, because the use of the forceps was not known at that time ; but, after Chamberleyn's discovery, this practice gradually sank in the estimation of obstetricians.

The thanks of the profession are due to Professor Simpson for the revival of this operation, and for the clear and lucid manner in which he has enforced his opinions both by argument and statistical data. He advocates turning in cases of slight distortion of the pelvis, and considers that the base of the head of an infant will pass with more facility and through a smaller aperture, when brought first, as in footling cases, than when it passes last, as in ordinary head-presentations. This doctrine, however, I ventured to differ from (see *Provincial Medical and Surgical Journal,* vol. ii. p. 3) ; and it has been more recently doubted by Dr. McClintock (*Obstetrical Trans-*

actions, vol. iv. p. 175). At that time, I considered perforation, if required after turning had been performed, would be
more difficult and more hazardous. This opinion has been
also lately expressed by Dr. McClintock.

Notwithstanding my former opinion, above referred to, and
the opinion I now hold, that turning in cases of protracted
labour, from the slighter contractions at the brim of the pelvis,
cannot ever become an alternative operation for the use of the
long forceps, there are, doubtless, some cases in which turning
would deserve a preference ; and, in the hands of some practitioners, it might be more safely undertaken.

The merit of the professor, in trying to establish this as an
alternative operation, is in the intention to abolish, as far as
possible, craniotomy.

"This practice of turning, in cases of pelvic deformity, is one
of the agitated questions of the present day, which requires the
sober and dispassionate consideration of all who are interested
in the establishment and advance of obstetrics."

This question can only be settled by a long accumulation of
practical facts and comparative trials. Mere opinion can make
no approach towards its settlement.

The danger of turning will be considerably diminished
when the plan of internal and external version, as recommended by Dr. Hicks, is adopted. (Vide *Obstet. Trans.,* vol. v.
p. 219 ; also his essay.)

In the performance of turning, I prefer and recommend the
operator to seize one foot or one knee, for reasons set forth in
my essays on various midwifery subjects. The funis is thereby
better defended, and the egress of the head is rendered safer
and easier by a partial breech-case having preceded it.

Dr. McClintock, " after having seized one leg and brought
it into the vagina, could not, with all the force he could use,
make the child revolve." He quotes the opinion of Madame
La Chapelle, who speaks of the difficulty of effecting version
by one leg when the head presents. She says the difficulty
is produced by the head being pushed into the brim before
the breech. It is quite evident that the cause which opposed
the revolution of the child was, not taking hold and bringing
down one leg, but the result of protracted labour after the
rupture of the membranes. The head was forcibly pushed

down upon or partially into the pelvis; and the uterus was doubtless violently, and perhaps spasmodically, contracted upon the infant's body, moulding and applying itself to all its hollows and projections.

As turning, in the cases of slighter distortion of the pelvis, is intended to save the child, this operation ought to be early performed—before, or as soon as possible after, the discharge of the liquor amnii. In cases which have been unduly procrastinated after this event has happened, and when the uterus is strongly embracing the infant, violent attempts to turn ought not to be made until some plan has been adopted to lessen the irritability of the uterus, and relax as far as possible its tonic and alternate contraction. Venesection and opiates are appropriate remedies. Does chloroform relax the uterus?

The death of the infant after the operation of turning (if it be living when this operation is commenced) is most frequently caused by the practitioner hurrying too rapidly its delivery after the revolution has been made. Time should be first given for the uterus to adjust itself to the changed position of the infant. When extractive force is used, it should at first be slow and gentle, and, if possible, in co-operation with uterine contraction. If the infant be rapidly and forcibly dragged through the pelvis, the chin leaves the breast, and is tilted upwards, thereby creating an unfavourable relative position between the diameters of the head and those of the pelvis. A great difficulty is now found to exist, which opposes an easy entrance of the head into the brim. Another mischief happens from attempts to draw the infant too quickly along by bringing the too bulky part of the infant (the head) to press upon and distend the os and cervix uteri before these parts are prepared to bear the change; and, consequently, spasmodic retention takes place, which is often so violent and obstinate as to cause the death of the infant.

III. *On Premature Labour.*—The induction of premature labour was first performed by Dr. Macaulay in England in the year 1756. But, although this is the fact, and though the importance of the operation is acknowledged and its adoption sanctioned by the most eminent British obstetricians, yet, if we consult statistics, we shall find it has only been limitedly

employed, in comparison with its practical importance. There
is no recognition of this practice in the statute-book, to distin-
guish it from that abuse of it which is committed for criminal
purposes. In this respect it is estimated in the same way as
some other operations which I shall consider in the subsequent
part of these remarks, being only sanctioned by the law of
custom. This is an unwise legislative omission, because it
permits wicked men to cover their crimes under the pretension
of a legitimate act. On this account, therefore, great caution
should be exercised whenever this operation is intended to be
performed. It is justifiable on moral grounds, and it is
approved of on every professional and social principle. The
object of its performance is noble and humane, as the lives of
those infants are saved by it which must otherwise be destroyed ;
whilst at the same time, according to my experience, the woman
incurs little risk, if any more, than that which is contingent on
ordinary labour, and very much less than that which results
from craniotomy. But, notwithstanding its high value, it
ought never to be performed without great necessity, nor
without having been first well considered and sanctioned by a
consultation.

It is a simple, safe, and efficacious operation, and, if duly
performed, infants not to be computed in number would be
born alive ; it saves when its alternative assuredly destroys.
It is not intended to supersede the Cæsarean section, for no
right-minded practitioner would ever think of adopting such a
course if the operation now under consideration were eligible ;
but the great object to keep in view is, to prevent as far as
possible the performance of craniotomy. It has, however, been
asserted by one writer " that premature labour should never
be attempted before it has been proved, by the event of one or
more destructive fœtal births, that the pelvis was so much
distorted that life must have been unavoidably sacrificed before
delivery could be accomplished, because a single fatal instance
is not always a sufficient warrant for the operation."

The destruction of one infant ought to satisfy every obstetrician that premature labour in a future pregnancy ought to
be induced, if the pelvis is so contracted that one full grown
cannot possibly pass through it. A second life ought never to
be sacrificed in such a case. But, in case such a pelvic

deformity were suspected, and, after a careful examination, proved to exist, it would be highly improper to allow a woman to go on to the full period of pregnancy, in order to prove by craniotomy the necessity for the induction of premature labour.

The longer gestation is allowed to proceed before the performance of this operation, a greater chance of living is given to the infant; but the precise period at which labour ought to be induced must entirely depend on the degree of positive or relative diminution in the pelvis.

The express object of the obstetrician is to save the infant; and, therefore, he should allow a sufficient length of time for its intrauterine existence, so as to enable it to support an extrauterine life. Most writers assert that it is not viable before the end of the seventh month of pregnancy; but I think it will live after a shorter sojourn in the uterus. Well authenticated cases are recorded of infants having lived who were born at six and at six and a half months. Similar cases have occurred in my own practice, and also in that of some of my British and foreign medical friends.

The following table shows the progressive development of the fœtal head which takes place during pregnancy, from the fifth and a half month up to the ninth :—

Date of pregnancy.	Bi-parietal diameter.	Occipito-frontal diameter.
At 5½ months	$2\frac{5}{12}$ inches	$3\frac{2}{12}$ inches
At 6 months	$2\frac{7}{12}$ inches	$3\frac{6}{12}$ inches
At 6½ months	$2\frac{9}{12}$ inches	$3\frac{9}{12}$ inches
At 7 months	$2\frac{10}{12}$ inches	$3\frac{1}{2}$ inches
At 7½ months	$2\frac{10}{12}$ inches	$3\frac{8}{12}$ inches
At 8 months	$3\frac{5}{12}$ inches	4 inches
At 9 months	$3\frac{1}{2}$ inches	$4\frac{2}{12}$ inches

The above measurements strikingly show that the size of the head will be found greater or less according to the period of pregnancy at which artificial labour is brought on; and the size of each must be relatively compared with the varying pelvic cavity through which any of them have to pass. The pelvis may be only just so much diminished in size as not to permit the passage of an infant at the seventh month, and yet allow an easy transmission of one at the sixth or even at the sixth and a half month. A very slight addition of bulk to the head, or a very little diminution of space in the pelvis, will mechanically oppose the passage of the infant; and, on the contrary, a very little comparative difference in either case will render its egress easy. This fact is exemplified by a very simple experiment. Add a little gold-beater's skin to a ball turned to the size or of the diameter of the space it would only just pass through, and this slight addition will be found sufficient to resist its passage.

In correspondence, then, with the facts just mentioned, I have ventured to recommend for consideration the induction of labour at the end of the sixth and the end of the sixth and a half months of pregnancy, in those cases in which the pelvic space is below that which is required for the safe delivery of an infant at the end of the seventh month.

The gradual development of the infant's head during pregnancy should never be forgotten when this operation is contemplated; and it is of equal importance accurately to measure the pelvis, in order that a correct relative comparison may be made. But if the calculation has been erroneous, and the head cannot pass unassisted, then, under such circumstances, the long forceps ought to be applied.

The value of the long forceps should not in all such cases be merely estimated as an accidental contingent auxiliary power; but, in some, this instrument from the first should be considered and accepted as an essential means for delivery.

The combination of these two operations will enable the obstetrician to safely extract a living infant which must otherwise be destroyed. This operation stands pre-eminently forward as conservative; and, as its mission is for the saving of both the lives, perhaps it may be said to be specially intended for the salvation of the infant which must otherwise be destroyed. We ought, in every case of difficult labour, whether terminated by forceps, or by turning, or by craniotomy, strictly and minutely to examine and ascertain the precise measurement of the pelvis, so as to be able to compute the advantages and disadvantages which are contingent on any of these modes of delivery, when compared to those of the induction of premature labour.

My remarks have been more particularly directed to the induction of premature labour in cases in which the pelvis is positively distorted. But it must be understood that, in all cases in which there is such a relative disproportion between the head of a full-grown infant and the available pelvic space, that the head cannot pass without a reduction of the size by craniotomy, while the pelvis is of such capacity as would permit one less (still viable) advanced to pass through at the above specified periods, premature labour ought to be induced; such as when the pelvis is regular in shape and symmetrical, yet too small from want of development; or when an exostosis has grown from some portion of it, or fixed solid tumours exist within it. It is also sometimes desirable to shorten the period of pregnancy in some diseases which threaten the life of the woman, as in some cases of albuminuria, or when violent uncontrollable vomiting exists.

If this operation be undertaken, subject to the restriction inculcated in these remarks, there can be no question as to the morality of the practice in such cases.

A very important object in medical jurisprudence is also gained by the practice of inducing premature labour at these specified periods of pregnancy. According to the English law, the descent of property is in some cases governed by the state of the infant when born. If it be living, or so far alive that the slightest vital movements could be perceived, such as a quivering of the lips, or the twinkling of the eyelids, then, under such circumstances, the husband would become entitled (by what is termed "the tenant by the courtesy of England") to the property. But if the infant be dead, then the right of inheritance passes in the line of consanguinity. A case illustrative of the above remarks occurred to Dr. Denman. (*Vide* Beck, and Dr. Paris and Fonblanque.)

Before the operation for the induction of premature labour is performed, the obstetrician should have fully acquainted himself with the relative measurements of the head of the infant and the pelvic space through which it has to pass. It is true that there is some variation in the size of the head; but these are exceptions, and the computation must be made on the average size. (See Table, page 43.) The development of the infant's head continues to increase during its sojourn in the uterus, and, therefore, the labour should be ended as near as possible to the time we had fixed for its completion, as there is some doubt how soon effective uterine contraction may ensue after means have been employed to induce it. It would, therefore, be most desirable to commence the operation a few days before the computed period for the fulfilment of labour; especially so when some of the measures are adopted for this purpose.

So long as the membranes are entire, and the infant is unrestricted in its movement and floating in the liquor amnii, its life is comparatively safe; but, as soon as the water is discharged, it is subject to more hazard, from the compression it must necessarily bear, and especially so if the os uteri be not dilated. If this be true in ordinary labour, it is more decidedly so when it is artificially induced. Besides, there is no other means so effectual in distending the cervix, and in dilating the os uteri, as the membranous bag filled with the liquor amnii, which acts during each pain as a powerful wedge.

If the infant should happen to lie in a bad position, and require turning, either the old operation or the internal and external manipulation (Dr. Hicks's method) could be undertaken with more ease to the practitioner, and with greater safety to the infant.

On these grounds, then, it is of the greatest importance, when labour is artificially brought on, that the membranes should, if possible, be kept entire; and, therefore, those means should be employed which are calculated to accomplish this object.

There are various measures proposed to induce premature labour, which I shall very briefly mention. Stimulant injections thrown into the rectum, and abdominal frictions, and the application of a firm bandage round the abdomen, &c., ought only to be considered as aiding others.

Secale cornutum, in repeated doses, has been prescribed; but, from its well-known poisonous effects on the infant, it ought never to be employed in these cases.

The old, and perhaps the most common, method of puncturing the membranes, is certain in its effects, although some days frequently elapse before labour ensues. Although I have formerly frequently adopted this plan, yet it is objectionable, on account of depriving the infant of the protective influence of the amnion fluid, &c. To partly obviate this evil, it has been proposed to carry an instrument through the os uteri and upwards between the uterus and the membranes, before piercing them.

Dr. Hamilton passed his finger through the os and upwards between the membranes and the uterus, and then round so as to detach them; and had the utmost confidence in it. His success was great. "Of forty-six infants thus prematurely brought into the world, forty-two were born alive." Although this method is safer than the older for the infant, several days sometimes elapse before uterine contraction is excited.

The vaginal douche has the confidence of many obstetricians. It consists of a forcible and continuous stream of water, sometimes warm and sometimes cold, being directed against the os uteri, so as to wash out the mucous plug. Although this is comparatively a safe measure, it is not always certain in its effects, and is also somewhat slow in acting.

The uterine douche, or injecting water into the uterine cavity by means of a syringe and an elastic tube passed between the membranes and uterus, has the approval of some practitioners. A considerable quantity of fluid has been thrown up by some; but this is a most hazardous experiment. Others have only injected three or four ounces. The risk of separating the placenta and throwing air into the uterine vessels renders the uterine douche rather objectionable.

Mechanical dilatation of the os by expanding instruments has been recommended; but such a plan ought never to be done; it is attended with great hazard.

Sponges, prepared so as to easily pass through the os, and left to expand, are comparatively safe, and may sometimes be employed as preparative measures. Distending the vagina with sponge-plugs, or by the introduction of a bladder which is afterwards filled by a syringe with water. One formed of caoutchouc is better adapted for the purpose. These latter methods are useful, precursory to other plans.

Elastic bags of various sizes have been contrived by Dr. Barnes, which are to be distended with water when they have been passed through the os uteri. As dilators, these contrivances are both safer and more efficacious.

The attempts to dilate the os uteri should be both gentle and gradual, and made to resemble as nearly as possible the

method Nature pursues in opening this part. Forcible dilatation, without preparation, is at all times most mischievous.
In cases of labour in which the hand has to be introduced
through the os uteri, this operation ought not to be undertaken
until this part (the os) becomes dilatable. So, in the operation of the induction of premature labour, our efforts ought to
be directed to attain, if possible, this state.

In December, 1844, I proposed galvanism as an important
means of arresting uterine hæmorrhage, and I also at the same
time recommended this agency for the induction of premature
labour; and my opinion still remains the same. If, however,
galvanism is not used to excite uterine action *de novo* in these
cases, its employment will be found most advantageous when
uterine contraction does not easily or vigorously respond to the
employment of some of the other measures. (*Provincial
Medical and Surgical Journal*, Dec. 1844.)

Whatever plan is adopted, we should never forget what has
been before said as to the necessity of having the parturient
process completed as nearly as possible within the period of
pregnancy fixed for its accomplishment.

CHAPTER VII.

THE MEANS OF DELIVERY BY DESTRUCTIVE OPERATIONS.

1. *The Induction of Abortion.*—The induction of abortion has been proposed for the purpose of superseding the necessity of the Cæsarean section; but in general, the woman has passed the period when it could be advantageously performed. In the great majority of such cases, she had arrived at the full period of pregnancy, and in many cases labour has actually commenced before the obstetrician has become acquainted with the malconformation of the pelvis. Sometimes he becomes acquainted with the pelvic deformity from the difficulty he has experienced in a former labour. It occasionally occurs, although very seldom, that the presumptive evidence of a malformed pelvis may be so strikingly observable in the early months of a first pregnancy as to induce the woman or her friends to apply to an obstetrician. But tumours of different kinds and exostosis may exist in the pelvis, without affording any indications whatever which might lead to a suspicion of their existence.

Whenever it is ascertained, after a most careful investigation in a first pregnancy, that the pelvic capacity is either positively or relatively too small to permit a viable infant to pass, then it would be justifiable, if possible, to perform this operation. But, if the woman again become pregnant, a question arises, whether it is justifiable again to adopt this plan. My opinion (which I submit with great deference to the profession) is, that it ought not a second time to be performed. If such a practice be admitted as sound, it establishes a principle totally at variance with the laws of God and man. The remarks to be made hereafter, on the comparative value of maternal and fœtal life will, I hope, be fully considered before such steps are a second time followed.

It has been remarked, that some teachers of midwifery in this country have asserted that this operation should be only once performed, after which the wretched woman should be left to take the fearful chance of the Cæsarean section. To say this, was to assume the position of the Supreme Judge. Just as well we ought to cure syphilis once only. It is said to be " a moral dogma, absurd and immoral, to prefer an ovum of four or five months, dependent for its existence on the mother." I shall not attempt to refute these remarks; but, after I have fully put my opinions before the profession, I shall content myself to leave the subject in their hands, with a desire that they will carefully weigh all the contingent circumstances, and consider whether there is any validity in my observations.

It is not on moral grounds alone that I object to the induction of abortion in order to supersede (as it is said) the Cæsarean section. It is not so safe an operation as it is usually represented. On the contrary, sometimes great danger has succeeded, and in some cases even death has ensued. Great difficulty has frequently been experienced in its performance, and in some cases it could not be accomplished. When the pelvis is highly distorted by mollities ossium, its entire character is changed; its cavity becomes altered in its shape, and all its diameters are very much diminished, whilst the depth anteriorly is very considerably increased. The position of the viscera which are normally contained within the pelvis is changed according to the degree of distortion. The uterus is especially altered in its relative position; it stands obliquely above the brim; and in most cases of extreme deformity, the os uteri cannot be felt. (See preceding remarks.) Under such circumstances, it is utterly impossible to perform this operation; and, in many cases, great risk is incurred by making the attempt. It is true that, by rash and rude manœuvres, so much mischief may be done that the expulsion of the ovum follows, although there has been no direct entrance into the os uteri, but solely in consequence of the injury inflicted upon the uterus or upon some contiguous organs, which is succeeded by great constitutional irritation, fever, &c., and sometimes by death. An experienced practitioner unsuccessfully attempted to destroy the ovum. The woman died from the effects. The pelvis is in my possession, and is an example of very great distortion from mollities ossium. Similar cases are also elsewhere recorded. In pelves distorted from the effects of rickets, in which the brim is elliptical, and the cavity and outlet comparatively more capacious, the difficulties above mentioned would be found considerably less.

I may be told that the difficulties and risks of attempting to induce abortion by passing an instrument through the os uteri, for the purpose of destroying the ovum, may be almost altogether avoided by the employment of the douche. From all the information which I have been able to obtain upon this subject, this measure has very frequently failed to produce any effect when employed in the early months of pregnancy.

II. *Craniotomy.*—Craniotomy is recognized in these king-doms as an operation of election, and is most extensively and frequently performed. It is employed in those cases of difficult labours in which the women cannot be delivered by the forceps, long or short, by the vectis, or by turning. It is also often had recourse to in many contingent accidents which happen during parturition, as in some cases of accidental and unavoid-able hæmorrhage, in some cases of convulsions, in some cases of rupture of the uterus, and also in those cases of protracted labours in which, from the neglect or ignorance of the practi-tioner, the pelvic organs and tissues are brought into such a state from pressure as to render delivery by other means hazardous to the life of the woman.

It has been proposed in cases of osseous deposits in the pelvis, on the grounds that it is impossible to estimate their density, and that most likely the structure would yield or even break down under the pressure made upon it during the extraction of the reduced head. This is, however, an unwise proposition, and ought not to be entertained upon such a presumption alone.

Now, when we contemplate the aggregate amount of infants destroyed by craniotomy in these countries for one year, the thought must be truly appalling. The facts of such a case cannot be, unfortunately, accurately arrived at. Reports of lying-in hospitals may, in some degree, show the force of this remark. Dr. Collins reports that, during his mastership at the Dublin Lying-in Hospital, 16,414 women were delivered, during which time craniotomy was performed in seventy-nine cases. Dr. Joseph Clarke reports that, in 10,387 cases of labour, craniotomy had been performed forty-nine times.

Now, assuming, in the first place, that craniotomy was performed relatively as frequently in the aggregate amount of labours in England and Wales which occurred in one year, as it was under Dr. Clarke's management in the aggregate of his cases, we should have 2,834½ infants annually destroyed by this operation ; and, by making a similar relative computation of Dr. Collins's cases, there would be 2,887¾ infants destroyed in one year. But, if a true statement could be obtained of the number of craniotomy operations which are annually performed in these kingdoms, the aggregate amount would far exceed either those of Dr. Clarke or those of Dr. Collins.

As it is so difficult—nay, I would say, quite impossible—to ascertain with arithmetical accuracy the real condition of the apertures and the cavity of the pelvis, we ought, in all cases of slighter degrees of distortion, as far as possible, to endeavour to save the infant, by first making a cautious and judicious trial of the forceps or turning, before we have recourse to craniotomy.

Then, as we have craniotomy performed in all cases in which the pelvic apertures or its cavity are either relatively or positively diminished, so as not to allow delivery by other means ; as we know that most of the obstructing causes to labours progressively increase in size—it must be evident that craniotomy must be much more difficult and dangerous to perform in some cases than it is in others. On this account I shall treat of it under two heads.

The first division includes those cases in which there is relatively no very great disproportion between the pelvic measurements and those of the head of the infant. In some of these cases, the mere perforation of the skull will suffice to set it free. In others, it will also be necessary to break up the brain, and then either partially or entirely to remove it from the cranial cavity ; after which it may be expelled by the uterus, or drawn out by the hand alone, or by the aid of the crochet, or by that of the craniotomy-forceps.

The second division embraces those cases in which there is relatively greater disproportion between the pelvis and the infant's head. There is, however, even in these cases, more difficulty experienced, and more danger attending the operation, in some than is found in others.

The pelvis may be distorted by malacosteon or rickets, or its cavity may be so much diminished by exostosis or tumours that a question may arise whether a mutilated child can pass through it. The opinions of different writers vary as to the space necessary to exist in the pelvis in order that a cranioto-mised infant can be dragged through it. Some state that a free space of one inch and a half in the antero-posterior diameter of the pelvis is quite sufficient; others say an inch and three-quarters is required; whilst others, again, affirm that, unless there be a space equal to two inches in this dia-meter, a mutilated infant cannot be drawn through the pelvis. One writer, however, has had the boldness to declare that he has delivered a woman when there was only one inch and a quarter space in the antero-posterior diameter.

These discrepancies of opinion as to the required space for crotchet-delivery are difficult to understand; but the different characters which pelves assume may in a measure account for them. But there has, doubtless, been some mistake made in either the accuracy of the measurement or in the age and development of the infant, as it is quite impossible to deliver a full-grown infant when such a contraction of the pelvis exists as the minimum above-mentioned. There is an unchangeable mechanical law which cannot be averted in these cases; that is, a body of a definite or given size cannot be drawn through an opening whose diameters are less than itself. Then, in order to reduce the infant's head, the vaulted part of it must be removed by taking away the two parietal and the frontal bones. Dr. Osborne reduced the head of the infant as far as possible, and then placed the base in (as he thought) the most favour-able position by turning it so as to bring it sideways first through the pelvis. He succeeded after great efforts, and delivered the woman, whose pelvis measured at the brim, in its antero-posterior diameter, only (as he says) one inch and a half. Dr. Osborne was doubtless mistaken in his conclusions, which has been so ably and clearly proved by Dr. Hull and Dr. A. Hamilton.

The measurement of the side of the base of the skull cor-responds with Dr. Osborne's estimation of it; but, in his

anxiety to astonish the profession, and to prove his great
achievement, he overlooked the fact that the other side of the
head had to follow; and that the bulk must, therefore, be
necessarily greatly increased by the addition of the cervical
vertebræ and the soft part of the infant's neck, which must
lie upon it and pass at the same time.

Now Dr. Hull has experimentally and indisputably proved
the fallacy of this assertion; and has shown that the least
measurement of the head, when it has been reduced to the
utmost, after dragging away the frontal and the two parietal
bones, is from the root of the nose to the chin; and, therefore, in order to bring the head (after craniotomy) through the
smallest possible pelvic space, the face must be brought first.
Dr. Hull, in Plate xii. Fig. 1 (*Observations*, &c.), gives a
representation of the reduced head, placed over a sketch of
the brim of a distorted pelvis; as well as the outline of that
of Eliz. Sherwood. So that, in such cases, it is most desirable
to convert the case from a vertex, if that were the original
presentation, to a face case. After this change has been
accomplished, the crotchet should be fixed in that situation, so
as to turn the chin towards the pubes; and the extractive
force should be directed so as to draw down and keep the chin
in the anterior part of the pelvis, as the natural flexion of the
head with the vertebræ facilitates the passage of the chin
under the arch of the pubes. Great care should be taken to
prevent injury being done to the uterus or the vagina by the
sharp edges or points of the bones of the skull, by placing
over them the scalp-integuments.

Dr. Davis says the necessity for Cæsarean section may be
reduced to zero by craniotomy; and, using his osteotomist, he
asserts that he has succeeded in bringing the reduced skull of
the infant through the space existing in the pelvis of Elizabeth
Thompson.

This operation was undertaken on a block of wood, in which
the pelvis was carved, &c. I had (now in St. Mary's Hospital)
similar blocks; but I could never accomplish the extraction.
However, it is one thing to operate on an inanimate machine,
however accurately formed, and another to operate in the
pelvis of a living woman. I had different preparations made

of the base of the cranium, with the vertebræ lying over the side and over the occiput, to show their relative measurement to the brims of different sized pelves, which were cut in wood.

But a practical question suggests itself: Will the performance of craniotomy meet all the difficulties arising from distortions of the pelvis? or can a woman be delivered by this operation, however greatly contracted the brim of the pelvis may be? Notwithstanding the great reduction which may be made in the head of the infant, and however favourably its base may be placed, I most unhesitatingly assert that it is sometimes utterly impossible, by any means whatever, either by the use of the osteotomist or of the cephalotribe, to deliver the woman. We ought never to disguise from ourselves that, in a great number of cases of extreme distortion, it is not possible to attempt delivery, as neither the os uteri nor the presentation of the infant can possibly be felt. This was found to be the case in many of the women whose cases are tabulated. (*Vide* Remarks on the Necessity.)

But, in many cases in which this information has been obtained, the head of the infant has been opened, and in some instances it has been reduced to its utmost limits, and yet the practitioner has been unable to drag it through the pelvis. Dr. Hull states in his "Defence" (p. 222), that ten women, upon whom embryulcia had been performed, lost their lives. He also relates three other cases, in which both the mothers and the infants perished. In one, neither the os uteri nor presentation could be reached; the uterus ruptured, and the infant escaped into the abdominal cavity. In another, after great difficulty, the breech was found to present. An attempt was made to pass the hand to lay hold of this part; but it failed, on account of the great pelvic contraction. After a second trial, the blunt hook was with great difficulty fixed; but, "notwithstanding all my exertions, the presentation could not be brought lower than the brim of the pelvis." She, therefore, died undelivered. In the third case, Ellen Gyte, who had been in labour sixty hours, the head was opened, and the crotchet applied; but, after the most strenuous exertions, it could not be brought through the pelvis. The vagina ruptured, and the infant escaped into the abdominal cavity. The pelvis was highly distorted ; the diameter of the largest circle that could be formed in the superior aperture was two inches and one-twelfth. The pelvis now belongs to me.

In a case in which the pelvis was highly distorted, under the care of the late Mr. R——, the head was perforated ; but, after the most powerful efforts, he was unable to bring it down. The uterus ruptured, and the child escaped into the abdomen. She died undelivered. A cast of the pelvis is in my possession. The cause of distortion in all these cases was mollities ossium. I have known several other cases in which the same melancholy events happened; and doubtless, if the grave could unfold the mysteries contained within it, very many more horrible terminations of pregnancy would be brought to light.

But, even granting that, in some extreme cases of distortion, there may exist sufficient pelvic space to permit an infant, whose head has been reduced to the utmost, to be dragged by great force through the pelvis, yet it must not be forgotten that such an operation must be extremely hazardous to the life of the mother. It is one thing to deliver the woman, and another to do so safely. It is much to be deplored that this operation is still permitted to be so unconditionally performed ; especially so when the injuries which are frequently inflicted on the pelvic organs, and when its comparative mortality are considered.

The statistics of craniotomy are in a very unsatisfactory state. After a very minute search, I have been quite unable to draw out such tables as I wished. Feeling strongly the importance of deductions which are to be drawn from an accumulation of well-authenticated facts, I published a letter (*Provincial Medical and Surgical Journal,* vol. xviii. p. 494, 1849) requesting that the Members of the Association would transmit me a statement of all the cases which had happened in each of their practices : but my appeal did not elicit much information. Not having been successful in this attempt, I could only avail myself of the statements already published by my worthy and esteemed friend Dr. Churchill, of Dublin, in his " Theory and Practice of Midwifery." The following is a copy of his tables :—

Authors.	No. of cases.	Mothers died.
Dr. Smellie	44	4
Mr. Perfect	3	0
Dr. Joseph Clarke......	49	16
Dr. Grauville	3	3
Dr. Ramsbotham	34	5
Dr. Maunsell	5	2
Mr. Gregory............	2	1
Dr. Collins	79	15
Dr. McClintock ...... Dr. Hardy	52	8
Dr. Beatty	3	0
Dr. Churchill	11	1
Mr. Warrington	1	0
	286	55

Or about 1 in 5.

" Independently of the abuse of this operation (craniotomy) —of its unjustifiable frequency—let us for a moment look at its relative fatality when compared with the Cæsarean section.

" According to the above" (Dr. Churchill's) " statistics," British practitioners resort to craniotomy once in 219 cases; the French, once in 1,205⅔ cases; the Germans, once in 1,944⅓. The average, therefore, of these three nations will be one in 896½. In 252 cases, 50 mothers died, or about one in every five. As regards the Cæsarean section, the same author states that he has collected 321 operations since 1750, from which 149 mothers recovered; and in 187 cases, where the result is mentioned, 130 children were saved, and 57 lost.

" Hence, then, we have a calculation showing that in craniotomy, where of necessity all the children must be sacrificed, one woman out of every five died; while, in the Cæsarean section, one mother recovered out of two and a fraction, and the success to the child was certainly most fortunate."

The destruction by craniotomy of a number of infants in different women, in successive labours, both in the practice of other obstetricians, as well as those which happened to myself; the ignorant and groundless adoption of this operation; the unprofessional and disgraceful manner in which I have known it performed (in one case the head was opened by a pair of scissors, which were obtained from some part of the family; in another case by a penknife); and the operation being frequently performed without a consultation—these circumstances, and deep reflection on the social and moral right to destroy life, convinced me that the present recognized practice in these cases ought to be modified.

In a course of lectures delivered to the members of the profession, I strongly denounced craniotomy as an operation of election; and I recommended it to be performed generally as an operation of necessity, and that it should only be conditionally accepted as one of election. These opinions I then expressed, and have ever since advocated and in every way promulgated. (*Provincial Medical and Surgical Journal*, and *British Obstetric Record*, &c.)

Such, then, was my proposition in 1843; and now, in 1865, after long reflection and matured experience, I am, if possible, more strengthened in my convictions. The remarks of Dr. Bedford, which I afterwards found in his translation of Chailly, bear so forcibly upon the subject, that I do not hesitate to quote them. He says :—

" The Cæsarean section is undoubtedly a dread alternative for the accoucheur to choose; but I cannot agree with Dr. Chailly, that its fatality is so great as he represents; nor am I disposed to adopt the opinion (unfortunately too general) that craniotomy is always to be preferred to the Cæsarean section. In truth, it needs some nerve, and, for a man of high moral feeling, much evidence as to the necessity of the operation, before he can bring himself to the perpetration of an act which requires, for his own peace of mind, the fullest justification. The man who would wantonly thrust an instrument of death into the brain of a living fœtus would not scruple, under the mantle of night, to use the stiletto of the assassin; yet how often has the fœtus been recklessly torn from its mother's womb piecemeal, and its fragments held up to the contemplation of the astonished and ignorant spectators as a testimony undoubted of the operator's skill. Oh ! could the grave speak, how eloquent, how momentous, how damning to the character of those who speculate in human life, would be its revelations !"

The facility of its performance has led to its abuse. Its recognition in the British obstetric code reflects no credit on the country, especially when its frequent performance is compared with the practice in France and Germany. It cannot be denied that craniotomy is a cruel operation; for surely no obstetrician ought to be so ignorant as to suppose that the infant *in utero* is void of sensibility. Yet there are some parties (if judged by their estimate of its life) who either professedly or actually believe it does not feel. This is, however, either a moral or a physiological fallacy; for there is not a doubt that it is endowed with this faculty in a very eminent degree, and consequently it must endure great bodily suffering from this practice. There are some practitioners who admit the existence of this principle, and, with the view of avoiding the infliction of pain caused by the perforator, delay opening the head until after its death. Craniotomy and embryotomy are the only operations which are recognized and justified by the British profession for the purpose of destroying life; but, although they are admitted into our obstetric code, they are not to be found in that of the law, and are only sanctioned by custom, and, through this usage, considered as "justifiable homicide."

There is no difficulty in understanding why so low an estimate of fœtal life is entertained when we consider what the doctrine is which is taught *in cathedrâ* and *extra cathedram*— " that, to save the life of the mother, it is justifiable to destroy the infant." From the early inculcation of this principle, the student becomes hardened to the performance of this dreadful task, and does it without compunction, and, no doubt, sometimes without reflection. My experience warrants me to make the above declaration, having met with many cases in which this operation has been most unnecessarily, and in some cases has been most unprofessionally, nay, most unjustifiably, performed.

Having now very fully expressed my objections to the recognition of craniotomy as an operation of election, I shall proceed to state my opinion in what cases it might be considered right to perform it.

When the infant is ascertained by the stethoscope to be dead, and the time for delivery has arrived, then craniotomy is justifiable ; but the labour ought never to proceed a moment longer after delivery is required in expectation of this event happening. The destruction of the infant from procrastination differs very little in principle from taking its life by the perforator, and, therefore, timely and other appropriate measures should be employed to prevent this event. Such are the long and short forceps, turning, and the Cæsarean section.

When some serious accident happens, such as rupture of the uterus, it would sometimes be admissible to perforate the head, and little compunction need be felt, as the infant is nearly, if not always, dead.

In a first labour in which the pelvic cavity is so diminished that a mature unmutilated infant cannot be delivered, and also in those cases in which this mischief has taken place after the woman has naturally borne one or more children, the operation may be performed.

Embryotomy is justifiable in cases in which turning is quite impracticable. In these cases the infant is nearly always dead.

In some cases of protracted labour, in which the pelvic organs or tissues have already sustained such great injury from pressure, and in which it would be extremely hazardous to the woman's life to deliver with the forceps, embryotomy may be practised. This event, however, will seldom or never happen in the hands of a judicious practitioner.

In some cases of hydrocephalic enlargement of the head, its size must be diminished by letting out the water by means of the perforator, or by a trocar. If this latter instrument can be successfully used, it is to be preferred.

CHAPTER VIII

I HAVE now to speak of two propositions intended to supersede the Cæsarean section, and which cannot be included under
either of the former divisions.

I. *Symphyseotomy.*—Symphyseotomy, or a division of the
cartilages constituting the symphysis pubis, was advocated by
Sigault, and his suggestion was received with enthusiastic
approval. A medal was struck off in honour of him.

British obstetricians have discountenanced this operation,
because it is not only inadequate to increase the diameters of
the pelvis, so as in any way to facilitate delivery when this
bony cavity is so contracted as to require the Cæsarean section,
but because it would be attended by most dangerous results.

British medical literature has only once been disgraced by
the record of the performance of this operation. I have
already adverted to Mr. Simmons's proposed compound operation of symphyseotomy and craniotomy.

II. *Mechanical Dilatation of the Pelvis.*—At a late discussion on a case of Cæsarean section at the Royal Medical
and Chirurgical Society, which was reported in the *Lancet,* it
was stated that a pelvis which was distorted from mollities
ossium might be dilated by means of bags introduced within
its cavity, and distended by either water or air. It was
asserted that this practice had been adopted in one case with
the effect of widening the pelvic space. The President stated
that he had found the bones affected with this disease yield
during the extraction of the child after craniotomy.

From my own practical knowledge I can truly affirm the
truth of the last statement. Some years ago I made the fact
known to the profession.

In a case at some distance from Manchester, in which the
pelvic space at the brim was about two inches and a quarter,
it was deemed right to craniotomise the child. After fixing

the crotchet, and adjusting the head in the most favourable position, force was cautiously used, and, after a few extractive efforts had been made, the head gradually descended, during which time the pelvic bones yielded to the pressure, and ultimately delivery was accomplished. Immediately afterwards the pelvis was examined, and found to have regained its former dimensions.

Other cases of this kind have come within my knowledge. One of great interest is briefly related in the *Provincial Medical and Surgical Journal*, vol. ii. page 706, 1847. Although it is true that the pelvic bones, when affected with mollities ossium, do sometimes yield to the pressure of the child when drawn through the cavity; and although the pelvis, as before mentioned, may be partially dilated by the mechanical influence of the elastic bag, yet, in my opinion, practical rules cannot be based on such an uncertain event where life is concerned.

Before such a change can be safely effected, a very considerable and an uniform softening must have taken place in the greater number, if not in all, the bones of the pelvis which are subjected to the influence of pressure, whether it be produced by the child or by artificial means. We know very well that this uniform change is found to happen in very few cases. Some of the bones sometimes become very soft, whilst others are comparatively unchanged. In other cases some of the bones become very soft, whilst others become very hard and brittle. Sometimes the pelvis becomes very highly distorted, and all its bones are extremely brittle and fragile, as happened in Dr. Murphy's case.

In a pelvis thus changed by disease, what would be the result (supposing it possible to accomplish it) of dragging a full-grown mutilated child through its cavity, or of attempting, by artificial and mechanical force, to dilate the pelvis for the purpose of accomplishing the delivery? The bones must be smashed, or at least so much broken, that irreparable injury must be produced. Even assuming that the bones are so uniformly softened as to yield to the pressure, it is quite certain that the increased capacity would only be temporary, as the bones would immediately and most likely completely return to their former position as soon as the pressure was removed.

CHAPTER IX.

THE British obstetric principle, which admits the preferential use of the crotchet, or the induction of abortion, is based on a calculation made as to the relative value of life of the mother and that of the infant or of the embryo.

It is said, and no doubt truly, that the social relations of the woman are greater than those of the infant. She is endeared to her husband, it may be to her children, and perhaps to her brothers and sisters, besides other kindred and friends. In the abstract these are weighty considerations, and are calculated to bring conviction to the mind, that her claims for the preservation of life greatly preponderate over those of the infant or of the embryo. It may be stated that these beings are unequal to the mother in organization, having no moral or religious responsibility, no social ties, no anticipation of their future doom ; and further, as regards the latter (the embryo), that it is at the very time drawing its nourishment from the mother's existence, that it has never had a distinct or separate life, and that it is little more than a member of the mother. These arguments, when only abstractedly considered, appear to be true ; and to warrant the deduction that the life of the infant or embryo is of little value when compared with that of the mother.

The impulse of natural feeling would probably—nay, nearly to a certainty—induce a man to decide in favour of this proposition. But, in the settlement of a question which involves the preservation or destruction of a human being, neither abstract reasoning nor feeling should be allowed to influence the obstetrician ;—conscience, reason, and judgment, ought to actuate him, after having fully and deliberately considered all the relative and contingent circumstances which either now or in future appertain to the case.

The unfounded and unwise opinions of Dr. Osborne primarily

and mainly led a large section of the profession to estimate
the life of the infant *in utero* at a very low value. He considered it nearly as a nonentity; as devoid of sensation, and
also as nearly deprived of motion. But I think I am asserting
the truth when I say that there are few, if any, members of
our profession who now entertain such opinions.

According to British practice, the destruction of the infant
is not limited to one; but if the cause which required its
sacrifice in the first instance be permanent, then in each successive labour, no matter what number, the same operation must
be performed. Hence in the end there must be a fearful
sacrifice of human life.

The repeated necessity of craniotomy in the same woman
demands from the obstetrician serious consideration. In some
cases, from one to twelve infants have been destroyed. Can
such a procedure be justifiable? The obstetrician should
pause; he could reflect. It is a dreadful position to be placed
in, to have one's hands imbrued with innocent blood.

The woman is *ipso facto* one party, and indeed the chief
party, who has brought into existence the innocent being whose
life the practitioner is employed to take away. It may be
argued, as a plea for her justification, that the wife is subject
to her husband; and there can be no doubt that she has
engaged to be so in the matrimonial contract, which was
mutual. But if it be considered right (which in such a case
as this could only be conditionally) strictly to observe this promise, it must be equally imperative upon both parties to obey
the law of nature and fulfil their mutual pledge to procreate
(and without doubt preserve) the species, both of which vows
are broken by the employment of the crotchet.

It may again be urged that both the parties were alike
ignorant of any cause (otherwise they were solemnly called
upon at the altar to avow it) which would interfere with the
great object of matrimony. Therefore, the woman, unacquainted
with her physical organic defect, would be entitled to have
those measures adopted for her first delivery, which would expose her life to the least hazard. Although the comparative
safety of delivery by the only two available methods is unsettled
by either positive or correct statistical evidence, yet, if she or
her husband desire that craniotomy should be performed,

then the obstetrician would probably act correctly in performing it.

But, in a second pregnancy, when they are fully acquainted that an unmutilated infant cannot be born, the question stands on very different social and moral grounds. The practitioner is here placed in a most responsible and trying position when called upon to decide whether he ought, time after time, or thus repeatedly, be made the agent to take away life. I entertain the fullest conviction that a great proportion of our profession have most conscientious scruples to repeat this revolting operation in the same woman. This destruction of infants, in my humble opinion, can be justified on no principle, and is only sanctioned by the dogma of the schools or by usage.

Dr. Denman had aversions to repeated crotchet-operations. He says: "Suppose, for instance, a woman, married, who was so unfortunately framed that she could not possibly bear a living child by any method hitherto known. The first time of her being in labour no reasonable man could hesitate to afford relief at the expense of the child. Even a second or a third trial might be justifiable to ascertain the fact of the impossibility." This eminent writer most decidedly erred in even conditionally sanctioning a repetition of this operation. In such cases, the impossibility of delivery of an unmutilated infant *per vias naturales* can and ought to be proved by a single case as clearly as by twenty, and when so shown, the Cæsarean section should be performed.

It is by no means to be understood that the life of the infant must never be sacrificed to save the mother. On the contrary, I have already enumerated cases in which craniotomy ought to be performed as an operation of election ; but it is not right to destroy the infant on the unfounded assumption that the mother could alone be saved by it ; a deduction altogether untrue, and unsanctioned by statistical evidence.

The life of the woman is not, either relatively or comparatively, always of the same value. If she be afflicted with a serious disease, or labouring under some incurable malady, being unfit and unable to discharge her domestic and her social duties, which performance can alone render her life desirable to herself or to her friends, then, under such circumstances,

F

the infant's life ought not to be sacrificed for the mere ideal chance of prolonging her miserable existence, which is a positive evil to herself.

Again, in our estimate of the comparative value of the two lives, we should especially consider whether the cause of difficulty is temporary only, or permanent. If it be of the latter character, then the infant's life (except as aforesaid conditionally) ought to be considered higher; and, if of the former kind, then we should invariably decide in favour of the mother. The obstetrician should, therefore, endeavour, as far as is compatible with the safety of the mother, to preserve the infant; for I know no case in which an intention or a desire to sacrifice her can ever be entertained, as the especial object of the practitioner should always be to try to save both lives.

When the contingent hazards of craniotomy and the risk of its abuse are considered, and as we know that the act is the sacrifice of life, and that this awful catastrophe must be often repeated in order to carry out the abstract proposition " to save the life of the woman by destroying the infant "—when we remember the difficulty which in extreme cases is experienced in performing it, the cruelty it inflicts, and many other evils consequent upon it—we may truly wonder that professional men should allow their minds to be haunted by an imaginary Cæsarean spectre, and be so obscured to their own moral and social responsibility. Why should the obstetrician stand in such an unenviable position, not only as an accessory, but *ipso facto* the agent? Again, I ask, ought he to be called upon, and ought he to consent, to victimize poor helpless infants in successive pregnancies, in numbers which make one shudder to recount? Does this remorseless sacrifice of human life correspond with those high moral principles which the members of our noble profession ought to possess?

The eminent Professor of Midwifery in Edinburgh makes the following pertinent remarks. He says: " Formerly, medical practitioners seem to have thought little, and medical writers said little, regarding the very repulsive and revolting character of the operation of craniotomy, when performed, as it frequently was, when the child was still living. Apparently some obstetric practitioners and writers of the present day continue to look upon the practice of craniotomy as one that should not unfrequently be adopted, and one which it is quite justifiable to adopt. Obstetric reports and collections of cases have been published within the last few years describing craniotomy as performed forty or fifty times, or oftener, by the hand of the same practitioner. But, perhaps, ere long it will become a question in professional ethics, whether a professional man is, under the name of a so-called operation, justified in deliberately destroying the life of a living human being."*

Woman naturally is mild, kind, and humane. She is endowed with great fortitude and undaunted courage. She has generally a great desire for offspring, and has a great love for children. Then how can we suppose that any woman with a well-regulated mind, if fully aware of her responsibility, could willingly be a consenting party to the repeated destruction of her unborn infants? According to my own knowledge the case is otherwise. I feel convinced in my own mind that there would scarcely be a woman to be found who would not suffer any amount of bodily pain to save her infant.

Every woman in whom there exists organic impediment to the passage of a mature or full-grown infant, ought to be at proper time fully informed of the nature and as to the degree of the obstacle. She should also be made acquainted with the alternative operations which are suitable to meet her case. If the obstruction be moderate in degree, then the forceps, turning, or the induction of premature labour, will be proper; but if these means are not available, or if the cause of difficulty is great in degree, then the performance of the Cæsarean section will be required.

* " Obstetric Memoirs and Contributions," vol. i. p. 606.

CHAPTER X.

In the preceding remarks I have to my own mind most satisfactorily proved that the Cæsarean section is at least an operation of necessity, and that those measures which have been proposed as substitutes are totally inadeqate to supersede it.

The (British) statistics of this operation are most certainly unfavourable; yet it has, I think, been shown that the great cause of the maternal mortality is avoidable, and that most of the other alleged causes of this fatality have been pointed out to be subject to control, and that some of them are really preventible, whilst others are remediable.

Although the deaths of the women from this operation, as hitherto performed, are very numerous, yet objections ought not to be raised against it on data so unsatisfactory as those are which now influence the opinions of British practitioners. When I speak of unsatisfactory data, I mean that we should not take an abstract view of them, and attach to them more importance than they deserve. British obstetricians have been guided more from prejudice arising from preconceived opinions than from an analysis of the real causes of death in these cases.

A comparative estimate of the mortality between this operation and craniotomy has never been fairly made. As regards the Cæsarean section, all the deaths are known, whereas those of craniotomy are only very partially known. This latter operation is not confined to any particular class of cases, but it is performed under very different circumstances and dangers —in some cases of accidents which occur in labours, such as hæmorrhage, convulsion, and other contingent mischief, which happen in women whose constitutional powers are unimpaired; whilst the greater part of the women who have undergone the Cæsarean section have laboured under incurable disease, and have had additional injury inflicted upon them by protraction, and in many cases by the practitioner in his ineffectual performance of craniotomy, &c.

Dr. Joseph Clarke reports forty-nine cases of craniotomy, in which sixteen mothers died, or one in three. Dr. Collins performed craniotomy in seventy-nine cases, in which fifteen died, or one in five. Now, in Dr. Clarke's practice collectively, there are recorded sixty-five deaths;* in Dr. Collins's practice collectively, there are recorded ninety-four deaths.* Thus the statistics of craniotomy, which are indiscriminately made up of all kinds of cases in which it has been performed, show an unfavourable result.

Besides this, we have no account of the injuries which are inflicted on the pelvic organs by the instrument used in this operation.

In seventy-seven cases of Cæsarean section there were collectively ninety-eight deaths,* the greater portion of which were not due to the operation, but, on the contrary, very many lives might have been saved if it had been timely and judiciously performed.

Notwithstanding all the pre-existing dangers in Cæsarean cases, several recoveries have taken place. These favourable terminations ought to encourage us to hope, and indeed ought to inspire us with confidence, that if the operation were earlier performed, and on a different class of subjects, it would be attended with infinitely more success.

These cases prove that, notwithstanding the serious nature of the constitutional disease which existed in these women, the vital powers were equal to the reparation of both the abdominal and uterine incisions, and also show the fallacy of the opinion that wounds of the uterus are necessarily mortal. The conservative vital powers were wonderfully apparent in the case of Mrs. Sankey, which is related in the *London Medical Gazette*, also in the *Provincial Medical and Surgical Journal*. The restorative powers in this individual were really so active as to impress my mind with the conviction that the chance of success would be as great in well-conducted Cæsarean cases as that which attends other capital operations.

Another woman (Mrs. Haigh), in whom mollities ossium existed to a great extent, and whose pelvis was very much

* These accounts contain the number of deaths of both women and infants.

distorted, showed great restorative powers.   She recovered,
and lived several years afterwards.   She died exhausted by
the disease.   *Post mortem* examination revealed no disease in
the abdominal or pelvic viscera.   The uterine tissue was uniform in appearance, there being no cicatrix to indicate the
site of the incision.   There was only a single band of lymph,
not thicker than a thread, passing from this organ to the peritoneum, so that there existed no mechanical obstacle to the
distension and ascent of the uterus if she had unfortunately
become pregnant again; but the moral rule of abstinence pre-
vailed with both her and her husband.

Recoveries after rupture of the uterus afford further evidence
that wounds of this organ are not always mortal.   The lacerated tissue in these cases must be in a very different, and
indeed in a much more unfavourable condition for uniting
than in Cæsarean cases.   In these accidents, the peritoneum,
the abdominal and the pelvic viscera, must inevitably sustain
very great injury by the escape of the infant, and also very
frequently of the placenta, through the uterine rent, and also
from the attempts which are made for the delivery of the
woman.   The same mischief cannot possibly be inflicted by a
well-conducted Cæsarean operation.

Two instances of recovery after rupture of the uterus have
occurred in my practice.   One of these women became pregnant several times afterwards.   In one of these pregnancies
she went to her full time and bore a child, which is still alive,
and she also aborted several times.   During her last labour,
and also during the several abortive periods, she had the
valuable aid and advice of my respected friend Mr. Hunt.
Many years afterwards she died in the Manchester Workhouse.
Her body was inspected by Mr. Hunt, in the presence of
Dr. Francis and of myself.   There was not the slightest trace
of the cicatrix in the womb to be seen; but there was a band
of slight adhesion to the ilium.

I was consulted by Dr. Clay in his first case of large ovarian
tumour, and attended along with him both before, during, and
after the operation.   I take this opportunity of saying I consider the successful issue of this operation as the commencement of a new era in the history of ovariotomy.   It had not
been attempted for many years before; and, at the time of its

performance, it did not stand as if it were a recognized surgical operation. I attended also along with Dr. Clay many of his next succeeding cases, being present at all the operations. I took great interest in these cases, not only on account of that which necessarily belonged to them, but also because the results analogically tended to substantiate my views relative to the probable success of the Cæsarean section. It is nevertheless true, that the influence in these two classes of cases is not quite the same; yet there is, however, sufficient similarity between them to lead us to trust more in abdominal surgery. In ovariotomy, there is certainly no uterine incision; but there is a necessary division to be made of the connecting tissue which exists between the tumour and the uterus. In many of these cases, extensive adhesions, which exist between the tumour and the peritoneum, &c., have to be separated. It is, however, evident that during the progressive development of an ovarian tumour, the sympathy of the peritoneum, &c., must in some measure be blunted, and consequently its susceptibility to mechanical injury must be diminished.

The rest of Dr. Clay's successful cases, and likewise those of Mr. Spencer Wells, and all those which have occurred to other practitioners, collectively afford strong evidence of the safety of abdominal incision.

The operations performed by my esteemed friend Dr. Blundell on animals, to prove some important physiological facts, likewise afford substantial evidence that abdominal wounds are very much safer than has been usually considered.

In my introductory remarks, I stated that I should not bring forward any statistical data, as shown from the result of foreign cases of the Cæsarean section, although I feel quite sure that the comparative position of this operation has been damaged by the omission. Continental success in this operation has been remarkably great, when compared with the results of British practice. There have been many instances of two or three successful operations on the same woman.

Having, as I sincerely hope, faithfully and candidly placed all the circumstances appertaining to the two operations—the Cæsarean section and craniotomy—before the profession, it now only remains for me to bring forward my proposition, first made in 1843. Deep reflection since that period, and a strong sense

of humanity, have induced me further to declare that the Cæsarean section should be generally performed as an operation of election ; and that craniotomy should be as far as possible abolished, and ought only to be performed as an operation of necessity, except (as already adverted to) in a very few cases.*
I am quite aware that many of the opinions I have so urgently stated in the foregoing remarks are at variance with those of the profession generally ; yet they have been most conscientiously advocated. They originated from the dictates of humanity, to try to extinguish as far as possible that dreadful expedient—nay, shall I not call it murderous operation ? —craniotomy.

* See remarks, pp. 59, 60.

PART II.

INTRODUCTORY REMARKS.

In the following observations it is my object to register all cases of Cæsarean section, published and unpublished, along with those contained in the first part of this work up to the present time, which (as far as I have been able to learn) have occurred in Great Britain and Ireland. I most respectfully refer my readers to the observations made in the First Part, as it is quite impossible, and, indeed, unnecessary, to enter at the present time upon all those subjects which are legitimately concerned in the consideration of this operation; as one of election, or even as one of necessity. The concentration of, or reference to, all cases of this important operation, especially as it is not universally adopted, or even recognized, is most valuable to those members of the profession who are desirous to analyze and compute the risk and dangers of it, in comparison with the results of those alternate operations which are recommended to supersede its performance, and thereby to satisfy themselves whether the unfortunate results of this (the Cæsarean section) are really attributable to it, or to other factors of mischief not necessarily belonging to it.

The annexed Tables are very limited when compared with those from which they have been extracted, and which embrace the following points :—the number and kind of previous labours; the mode of delivery; the tangible state of the os uteri; the situation of the placenta; the line of the incision; the previous and present constitutional state; the administration of chloroform; the application of ether spray; the reputed cause of death; and the results of the *post mortem* examination.

I have enumerated the above points, which head the columns

in the said Tables, in order that the reader may understand upon what plan I have based my deductions.

I have had a strict analogical investigation made of the statements contained in all the cases tabulated from No. 77, as those up to and including 131.

All important practical matters are recorded in the following named sections—viz., statistics, maternal mortality, infantile mortality, exhaustion, peritonitis, hæmorrhage, successful cases, chloroform, ether spray, sutures, operation.

Twenty-one of the cases now tabulated, were so tabulated and published in my farther observations—viz., the cases numbered from 77 to 98 inclusive, and the cases now for the first time published are 33, named and numbered chronologically from 99 to 131 inclusive, there are therefore now 54 cases registered in the following tables, of which number 45 occurred in England, 6 in Scotland, and 3 in Ireland. I may again remark as I did in the First Part, page 4, that it is a remarkable fact there is no case recorded from Wales. The causes which have rendered this operation necessary in these 54 cases are as follows :—in 8 cases the pelvis was distorted from *mollities ossium ;* in 17 it was caused by rickets ; in 1 the pelvic bones were undeveloped, and the apertures diminished— characteristically partaking of the oblique shape described by Naegelé, only differing in having both sides although not equally affected ; in 4 cases the obstacle was caused by pelvic distortion—the character of the distortion is not mentioned ; in 2 large exostoses, in one of which it sprang from the upper portion of the sacrum which caused the difficulty ; in 2 it was produced by large fibrous tumours blocking up the pelvic cavity ; in 9 cases carcinomatous degeneration of the cervix and os uteri ; in 1 case carcinoma of the rectum rendered the operation necessary; in 1 case the operation was performed on account of a large hydrated cyst, which had passed from the abdomen into the pelvis and was placed before and compressed by the advancing head of the infant—it was so hard as to be mistaken for a bony growth. In 1 case the obstruction was produced by induration of the vagina ; in 1 case the obstruction was produced by the injuries which the pelvis had sustained by falling down stairs in early life.

Although the diminution of pelvic apertures has been

deemed sufficient in these cases to prevent the delivery per *rias naturales,* that it must ever be borne in mind that the cause of mischief in most of them is not stationary but progressive, so that ultimately the pelvis might be as nearly as possible blocked up.

In Case 79, in which the hydrated cyst existed, the error of diagnosis which was made was proved by the *post mortem* examination. It might have been obviated by an earlier examination before the cyst had been so firmly compressed. Puncture of it at an early stage of labour would doubtless have removed the cause of obstruction.

In some of the cases the os uteri could not be felt, whilst in others it could only be reached with great difficulty. There is ample evidence in all (except the one just alluded to) of the cases to prove the necessity of the Cæsarean section. The mechanical obstruction which existed is sufficient to demand its performance; but the unsuccessful attempts made in some of them by some obstetricians to deliver by craniotomy, embryotomy, &c., afford further corroborative evidence that this operation cannot be eliminated, but it must stand at least as one of necessity, which neither craniotomy, cephalotripsy, the induction of premature labour, or of abortion can supersede.

There is one case, No. 97, stands in the history of this operation quite unique. There are numerous important points belonging to it, and although I have analytically brought some of them before my readers, yet the whole ought to be read and studied by my readers as cited in the *Medical Times and Gazette,* vol. ii. 1865, page 359.

Mr. J. Spencer Wells commenced his operation solely with the object of removing a large ovarian tumour. Before he had completed it, and cut through the pedicle, he found another tumour which he considered to be the right ovary cystically enlarged. He now passed a trocar into it, and after withdrawing it, three pints of a bloody fluid escaped through the canula. The tension being thereby so much lessened he was now able to see the Fallopian tube passing from the upper part, which at once convinced him that he had punctured the gravid uterus. On the canula being withdrawn, a soft, spongy, bloody mass protruded. He now introduced his finger in order to push this back into the uterus, and also to examine

the cavity of this organ, during which its tissue, which was
soft and friable, as if it had undergone fatty degeneration, gave
way along the middle from the puncture (which was near the
fundus) to the extent of 3 or 4 inches towards the neck.   By
very slight pressure a quantity of liquor amnii, and also a fœtus
of five months escaped.

The foregoing circumstances constitute what may be considered the cause or necessity for Cæsarean section.   The clear,
bold, and unequivocal statement made by Mr. Wells reflects
the highest honour upon him, and his conduct throughout in
relation to the case affords a noble example of high professional
morality.

Dr. R. P. Harris, of Philadelphia, who has done so much
to establish the operation of Cæsarean section as one of election by the successful results of his practice, in his communication to the *British Medical Journal*, No. 1,005, April 3rd,
1880, says :—

"It may appear strange that an American writer should
undertake the continuation of an English Cæsarean record ;
but the fact is that the statistics were collected as a matter of
necessity in the prosecution of another work, and not with
any view to their being published abroad.   They having been
collected, it occurred to me that the Table, which I have no
intention of publishing here, should be presented for this purpose to the *Journal* which contained, in 1868, the last tabular
statement of Dr. Radford, amounting to a record of twenty-
one cases.

"After having tried in vain, on several occasions, to induce
obstetrical workers in London and elsewhere to collect the
Cæsarean statistics of the last ten years, I was obliged, in
order to make use of the record, to search out the cases
myself and arrange them in tabular form.   Having done this,
I discovered some points of interest that I had not anticipated,
and which must be new to many in Great Britain who have
not examined thoroughly the records of mortality under this
operation, which has been so much more fatal there than with
us ; and, for this reason, I have sent the Table to England,
where it will be of much more interest than it would be in
America."

I believe Dr. Harris applied to me ; but at that time I did

not feel myself equal to undertake the work, and therefore I declined to do so.

The cases recorded in Dr. Harris's Table are contained in mine, and are numbered so as to stand in chronological order. Dr. Harris has made some valuable deductions which are well worth perusal. He has been most indefatigable in his exertions in searching for all unpublished cases in the United States. He has been very successful in obtaining an immense number of cases which would have remained unknown but for his persevering energy.

In the desire to achieve a complete knowledge of the number of cases which have occurred, and the causes why so many perished; why they did so and how we can make the proportion less, and what would be the probable result if the operation had been performed at a very early stage of labour, or under different conditions in the physical health and habits of the subjects—on whom does the responsibility rest?

In Great Britain and Ireland the fault of delay rests with the great bulk of obstetricians, whose minds are erroneously biased against the operation. No doubt the previous bad state of health, the depraved organism from mollities ossium, has a pernicious influence on the result of the operation; but this latter named cause no doubt will also influence the results of craniotomy—an early operation is mostly important. (*See* Part I., page 24; also Section on "Operation.")

The record of the cases of women who recovered as to the time they were in labour showed a result which would rather weigh against the general opinion of protraction being the most important factor of mischief; and notwithstanding the evidence is as it is stated, yet I am convinced that delay is, as is usually considered, one of the great sources of mischief. In those cases of recovery in which there has been great delay, there must exist in those cases factors which act favourably on the vital powers. Dr. Harris says: "The great factors of the disqualifying condition I believe to be poverty : the want of proper nutrition; the existence of malacosteon ; and superadded to these the habitual drinking of beer or gin."

There can be no doubt that where any of the three first-mentioned conditions exist, either singly or collectively, the woman will be in an unfavourable state to undergo so important an operation; but as to the habitual drinking, my experience warrants me in declaring I have never known this practice to exist in those women at whose operations I have been present and concerned in their performance.

In the United States no case of cancer or malacosteon has been operated upon, but rachitic subjects have constituted one-half of the cases. It is very evident that the Cæsarean operation is much less fatal in women in America than it is in Great Britain and Ireland, under whatever circumstances the operation may be performed.

I will just quote the termination of a letter containing the statistics from Dr. Harris, which I will insert as showing the great success in his operations. " I will stake my own work against the world, as I believe it fully represents the bad and unreported work of the country. We have had one hundred and thirteen operations and forty-nine cures in the United States; one hundred and twenty-one operations and fifty-six cures in the whole of North America. Twenty-eight early operations in the United States saved twenty-one women and nineteen children. Twenty-three children were delivered alive. Nine women in North America and eight in the United States were operated on twice, with sixteen recoveries in the eighteen operations—fourteen children saved. Nineteen cases of uterine suture with six recoveries; of three early cases two were saved; of sixteen late cases four recovered.

" 65 published cases ... 36 women saved.
48 unpublished „ ... 13 „ „
—— ——
113 49
Published cases saved... 54 1⁄13 per cent.
Unpublished „ ... 27 1⁄12 per cent."

I beg to present my thanks to all those gentlemen who have so kindly responded to my request in furnishing me with unpublished cases of Cæsarean section.

And I also thank Dr. Walter, the obstetric surgeon of St. Mary's Hospital, Manchester, for the assistance he has rendered me in preparing the Table of unpublished cases of Cæsarean section as seen in Table 3.

ON CRANIOTOMY AND CÆSAREAN SECTION.

THE relative position which craniotomy and the Cæsarean section should hold in obstetricy is unsettled. There are some obstetricists—no doubt the great majority—who consider that craniotomy should stand as an operation of election, and the Cæsarean section as one of necessity. Others think differently. There is no law to bind practitioners to adopt either one course or the other, but the decision is left to the judgment of the medical attendant of the case.

Some have expressed opinions as to the practice to be adopted in cases of difficult labour from extreme distortion of the pelvis, in which either craniotomy or Cæsarean section may be required, which are so much at variance with those I entertain that I am induced to make a few comments.

It would be well if moral laws governed medical men in all their professional undertakings, not only in reference to the Cæsarean section, but especially so when craniotomy is contemplated. No man would preferentially adopt the Cæsarean section because he was an expert, or because he was " dazzled," perhaps, by its false brilliancy as an operation deserving to be raised into competition with turning, craniotomy, or cephalotripsy. Any man undertaking such an important operation with any such motive must be deficient in both morality and common sense.

Cæsarean section has been denounced as " too often sacrificial in fact," although " conservative in its design." An unjust or wanton sacrifice of life is criminal, but how the death of a woman, happening after the performance of an operation legally permitted and sanctioned by medical ethics, and undertaken for the preservation of the lives of both mother and infant if then living can be placed in such a category, I am at a loss to understand.

The Cæsarean section is strictly conservative, for what can entitle an operation more to such an appellation than the

G

salvation of nearly all the infants which are living at the time when it is commenced?

And if the operation was performed early, before the influence of protraction was produced, I have not the least doubt but many more would have been preserved. (See section on Infantile Mortality.)

Statistics, as put forward to prove the necessity of any operation, may not be altogether reliable, but there is not a chance of greater fallacy existing in respect to the Cæsarean section than would be found in relation to craniotomy or to any other operation.

In Great Britain and Ireland, mothers saved are only comparatively few, being only 17·55 per cent., but if the causes of mortality were strictly sought for, it would be considered a high percentage.

The operation has not, in the great majority of cases, been performed before the woman has been nearly brought to the brink of the grave. If all cases in which fatal morbid effects stamped upon the woman by neglected protracted labour and other factors were excluded from the calculation, the result would show a very much more favourable aspect.

In the First Part of this work, page 70, I adverted to the recovery from the operations of ovariotomy, but since the date of this statement, an immense number of women have been saved by it.

In the report of the practice it was stated the other day in the *British Medical Journal*, June 19th, 1880, that Mr. Spencer Wells, had, June 11th, performed his one thousandth ovariotomy operation.

The reports of the practice of Dr. Keith, Dr. Clay, Dr. Lloyd Roberts, and that of a number of other operators, ought to lead us to expect that, if the same antiseptic treatment, and if other plans which are so strictly carried out as in ovariotomy cases, were adopted in Cæsarean operations, we should find an immense increase in the number of women saved.

The recovery of several women with extra-uterine pregnancy upon whom abdominal section was performed and precautionary treatment was adopted at St. Mary's Hospital, Manchester, affords further encouragement for us to expect a more successful issue in Cæsarean cases.

The successful treatment of cases of rupture of the uterus in which the child had passed from the uterus into the abdomen, prompt a hope that Cæsarean section will be ultimately estimated as an operation of election.

In another case of rupture of the uterus, the whole body of the child had escaped from the uterus into the abdominal cavity, except the head, which was retained at the fundus; an abdominal incision was made, the child and the placenta were extracted, afterwards the integuments were closed, the woman recovered, and, I believe, afterwards became again pregnant, and had a propitious labour. This case affords another illustration of the wonderful curative powers of Nature and what violence the peritoneum will sustain. " Porro's operation has been performed thirty-six times. In five of these the operations were performed under conditions which made recovery hopeless, but of the remaining thirty-one eighteen recovered."

This is a favourable result. Fifty-eight per cent.

This operation is much more serious than the Cæsarean section. It includes the abdominal and uterine incisions in order to extract the child and placenta, and in addition the entire uterus is afterwards extirpated. I refer to this operation as another example of the tolerance of the peritoneum to have great injury inflicted upon it and recover.

Dufeillay declared " that the operation performed under favourable circumstances as early as the improbability of delivery *per vias naturales* is recognized, gives nearly seventy-five per cent. of recoveries."

There is no ground whatever to doubt the truth of this author's statement. It is impossible to have a correct estimate of the fatality which really belongs to either the Cæsarean section or to craniotomy so long as the operation is delayed until irretrievable mischief is inflicted.

Craniotomy is most assuredly a sacrificial operation in its designs and in fact.

I have in a former communication (as I do at the present time affirm) stated that we have no authentic registry of fatal cases consequent on this operation (of craniotomy) when performed in cases in which the pelvic apertures are very much diminished.

Dr. Churchill states that one woman in five and a half (or

twenty-seven per cent.) dies; his Table contains all kinds of operations. If the Cæsarean section is only accepted as an operation of necessity, it is very important for its advocates to know, if possible, the exact degree of pelvic contraction which limits delivery *per vias naturales.*

There is considerable difference of opinion between different writers as to the space required to drag a craniotomized full-grown infant through the pelvis.

Dr. Osborne has said he delivered a woman by craniotomy whose pelvis only measured one and three-quarter inches in its antero-posterior diameter, having removed the parietal and frontal bones, and afterwards having turned the base of the skull edgeways. Now this is a physical impossibility, which I have practically proved. For a full explanation of this question see Part I. page 53 ; see sketch of the brim of the pelvis of Elizabeth Sherwood, Figure 10.

Dr. Hull, as I have before said (see *Medical Times and Gazette,* November 21st), ... has also proved Dr. Osborne's fallacy.

Dr. A. Hamilton has also most unquestionably refuted the statement of Dr. Osborne (see his letters to this writer), and says it is impossible to deliver *per vias naturales* unless there is a greater pelvic space. Dr. Burns says Dr. Osborne has been deceived in his estimate of the pelvic conjugate diameter in his case, and says that a full-grown infant cannot be delivered by craniotomy unless the short pelvic diameter fully measures one and three-quarter inches. Dr. Kellie says he delivered one woman whose pelvis measured in the antero-posterior diameter one and eleven-sixteenths, on one side two and one-sixteenth, and on the other one and a half inches. He delivered another woman whose pelvis measured in the antero-posterior diameter only two inches. Dr. D. Davies says, if the intermediate space between the pubes and the promontory of the sacrum be no more than two inches, there would be the greatest risk of inflicting such injury to the maternal structure as to lead to the death of the woman. By the use of his osteotomist his estimate of the pelvic space required is considerably less, which I shall particularly refer to when I speak of instruments which are used in craniotomy.

Dr. Churchill states two inches at least are required to exist in the pelvic conjugate diameter for delivery of a full-grown infant by craniotomy.

Besides those authorities above enumerated who define the minimum space through which a mutilated infant can be brought, I will quote several others who emphatically assert that, if the transverse and conjugate diameters measure as stated as follows, delivery can be effected *per vias naturales* :—

1 Barlow	3 × 1½		5 Hamilton	3 × 1½
2 Barnes	3 × 1¾ to 1		6 Leishman	3 × 1¾ to 1½
3 Burns	3 × 1¾		7 Ramsbotham	3 × 1½
4 Campbell	3 × 2		8 Playfair	3 × 1½

Doubtless the above-named practitioners are counted as of the highest eminence in the obstetric department of the profession ; but notwithstanding they are so, I most conscientiously differ with these assertions, and consider them as too dogmatic and unconditional. The data on which they are founded, I presume, are derived from pelves distorted by rickets, in which the cavity and outlet are more capacious than exist in those distorted from malacosteon, or pelves in which the cavity is fully occupied by a large exostosis, or a large fibrous tumour. (See article on "Craniotomy" in Part I. page 51).

Dr. Leishman says (see "System of Midwifery," page 613, after stating that craniotomy may be successfully performed in contractions of one inch and three-quarters, and the experience of some modern operators state this limit may be reduced to one inch and a half): "We may say, then, confidently, that when the conjugate diameter exceeds these limits, we are in no case justified in at once deciding in favour of the Cæsarean operation. We must once more, however, reiterate a former observation, and call attention to the fact that the conjugate measurement is not alone to be taken into account as it is too much the fashion to do, seeing that we may have irregular or angular distortion, in which the other diameters are similarly, or it may be chiefly, distorted. What we wish, therefore, now particularly to notice is that the conjugate measurement cannot be accepted as the test of the necessity which may be assumed to exist for the performance of this operation."

At the meeting of the members of the Dublin Obstetrical

Society, May 1st, 1880, a discussion took place on Dr. Kinkead's Paper on " Craniotomy and its Alternatives : Cæsarean Section," &c. &c. Several of the most eminent obstetricians in Dublin took part in the discussion, and therefore I consider it right to quote some of their opinions.

Dr. Kidd, Master and Obstetric Surgeon, Coombe Lying-in Hospital,&c., in speaking of slighter degrees of diminution of the pelvic conjugate diameter, says : " Authors variously mention antero-posterior diameters of from 3½ inches to 3 inches or less as the smallest through which a living child may pass. At the bedside, I believe, this difficulty can never arise where you can have opportunities of comparing the size of the head lying above the pelvis with the size of the pelvis itself, and can apply a forceps carefully once or twice, besides having the assistance of a person in whose judgment you have confidence. If we are once satisfied, after due and careful trial, that we cannot bring the head through in an unmutilated condition, and that delivery can be easily and safely effected by the operation of craniotomy, I maintain that it is our duty to lessen the head, and to deliver the woman."

Presuming that it is the first time such cases as those just referred to occur, I, with great readiness, agree with the practice suggested, being satisfied, at the same time, that the induction of premature labour would meet the difficulty, and prevent a second sacrifice of a child. In coming now to the question of the danger in cases of extreme narrowing of the pelvis, Dr. Kidd states these cases very rarely occur. He says : " Early in my connection with Coombe Hospital, I came to the determination, that if ever I met with a case of this kind I would very carefully consider the propriety of performing the Cæsarean section instead of proceeding to craniotomy. I believe the question of craniotomy or Cæsarean section to be one that must be decided by each man's own experience, and that it cannot be decided by any statistics that have been produced. If I could deliver a woman with a cephalotribe, I should be sorry to expose her to Cæsarean section in any of its forms." After speaking of American statistics, " I do not accept the proposition that, even with so narrow an antero-posterior diameter as two inches, it is our duty without connecting anything else, to perform Cæsarean section. If a tumour occupies the brim of the pelvis which you cannot push away or diminish in size by tapping, the question of Cæsarean section or craniotomy very fairly arises. I believe Cæsarean section would probably afford the mother a better chance than dragging the child through the pelvis, lacerating and bruising the tumour, and perhaps setting up inflammatory action." In a case of malignant disease of the uterus, with the cervix greatly thickened and hardened, and perhaps a scirrhous mass developed in it, the woman would be more easily delivered, and would have her life prolonged, by the Cæsarean section than by craniotomy. But in case of cephalomatous disease, with no great induration or surrounding infiltration of the tissues, delivery by craniotomy would be safer than Cæsarean section.

Dr. McClintock appears to entertain the same opinion expressed by Dr. Kidd, both as to rarity of cases of extreme distortion of the pelvis and as to the practice to be adopted. He says : " So far as I can form an opinion without having had direct experience, I agree with Dr. Kinkead that craniotomy, in cases of extreme pelvic deformity, has been attended with nearly as large a rate of mortality as Cæsarean section performed early in labour."

These two obstetricians do not agree as to Porro's operation.

Dr. E. B. Sinclair, King's Professor of Midwifery, School of Physic, Trinity College, Dublin, after having stated his opinions as to the operation of craniotomy in cases of slighter narrowing of the pelvis, says : " But when we come to cases of extreme narrowing, where craniotomy cannot be performed without lacerating the parts, and where we find, from examination, that the operation would be so seriously dangerous to the woman that in all probability she would die under or after it, then Cæsarean section ought to be performed in preference to craniotomy.

" With regard to mortality from craniotomy, the fair way to state it is this : In the first two classes of cases I have mentioned the mortality from craniotomy is almost *nil*, while in cases of extreme narrowing it is not even one to four, but one to two. These are cases where Cæsarean section comes in as an operation of selection.

" If, as in the case of the unfortunate woman mentioned by Dr. McClintock, the narrowing is of such a degree as 2½ or

2 inches, craniotomy would be obviously almost certain death to her. Would she not have a better chance of life with Cæsarean section? As good a one almost as a woman who is cut for dropsy of the ovary. I think that in all cases where extreme narrowing exists, Cæsarean section should be an operation of election, and not one of *dernier ressort*. I have taught that in my class for years, and if I were to have such a case in my extern maternity, I should bring the woman into the hospital, and at once, if permitted, perform Cæsarean section. Where you have a narrowing of the pelvis coming to 2 inches, I say that craniotomy is fraught with so great a danger as involving almost certain death to the woman, and in such a case you should have no hesitation in performing Cæsarean section *at once :* you should not procrastinate, but perform the operation the moment her labour sets in. If you do you will doubtless have success. In one case of great narrowing which Dr. Johnston and I have recorded, craniotomy was, after consultation, performed; the parts were lacerated, and immediate death was the result. If Cæsarean section had been performed at once, we should probably have saved her life and that of her child. Of course, when the child is known to be dead, craniotomy may be performed without hesitation; but it is a terrible thing to have to kill the child in order to save the mother, though such a contingency rarely occurs.

" In my opinion, when there exists extreme deformity, your best chance of saving both mother and child is to perform Cæsarean section as soon as labour sets in.

" It is fortunate, however, that in this country we have so few cases of deformity. I believe, as I have said, the reason why we have failed in this country with Cæsarean section, is because we have delayed the operation instead of performing it at once."

I have been induced to largely quote the opinion of the foremost and eminent obstetricists who took part in the discussion on Dr. Kinkead's Paper, because it shows what the Dublin Obstetricy is in reference to the opinions entertained as to the Cæsarean section and to craniotomy.

Some writers who advocate and sanction craniotomy in cases of extreme pelvic distortion have never performed the operation, nor ever even witnessed a labour which was thus obstructed. Some of them, however, have the candour to acknowledge their want of personal experience. Although it is so positively stated that a mutilated full-grown infant can be brought through the minimum of pelvic space, can we be assured that the operation can be safely performed? And is there not quite as much danger to the life of the woman as is incurred by the Cæsarean section? I have not the least doubt there is quite as much risk to the life of the mother from craniotomy performed in the higher degrees of pelvic distortion as there would be from Cæsarean section, both operations being freed from the contingent mischief produced by protracted labour.

Writers of great eminence entertain the same opinion.

Can the dimensions of the pelvic apertures be ascertained with such mathematical accuracy as to justify the practitioner in having recourse to craniotomy in the more extreme cases of pelvic distortions? From my own long and extensive practical experience I am firmly convinced that it is impossible to ascertain with precision the exact degree of pelvic contraction in such cases as those in which a very small fractional mistake makes the difference between the life and death of the woman. Even in the lighter shades of pelvic distortion it is quite impossible to compute the exact measurements.

My opinions are substantiated by our best writers and able practitioners. Practical results also corroborate the above statement.

Many cases are recorded in which the head of the infant has been opened by obstetricians of high character, but who have afterwards been unable to deliver the woman; the patients have been left to die after the infants had escaped into the abdomen through a laceration in the uterus. I have met with cases of this kind.

One is recorded in the "London Obstetrical Transactions," vol. viii. p. 158. In some other cases abdominal section has been performed in order to extract the mutilated infant.

There are several cases among those tabulated in which the head of the child has been opened, and the pelvic structures subjected to all the mechanical injuries which necessarily attend the attempts which are made to bring a mutilated child through a very contracted pelvis.

After unavailing efforts to drag it away, the poor creature

has been doomed to undergo the Cæsarean section in order
that she may be delivered, and then die; an event sure to
happen, and then her death must be placed to the mortality
of this section.

It is stated if there be a space in the antero-posterior conjugate diameter of 1¾ inch, and transverse diameter equal
to 3 inches, the base of the head of a full-grown infant face
brought first can be dragged through the pelvis. I deny the
truth of this statement, unconditionally put forth as a general
rule, regardless of the kind of distortion of the parts of the
pelvis involved in the mischief. The comparative difficulty of
performing craniotomy, and the issue to the mother, are in a
great measure dependent on the characteristic differences
which exist in the distorted or contracted pelvis. In the
rickety pelvis, or in those in which the brim is ellipsoid or
oval, varying in degree of contraction in different parts of the
opening, craniotomy can be performed with greater care and
under higher relative degrees of contraction, because the
outlet is more capacious; but in those in which the pelvis is
distorted by malacosteon, and all the parts of the pelvis, brim,
cavity, outlet, depth, &c., are more or less affected, there is
more difficulty and more danger from the operation.

In many of this kind of cases the antero-posterior diameter
is considerably greater than the measurement fixed by cranio-
tomists as ample for the base of the skull of a full grown
infant to be brought through, and yet it would be found quite
impossible to do so. There is practically more space than is
said to be required, and yet relatively there is less.

As an example, let us take the case already referred to in
the "Obstetrical Transactions," vol. viii. p. 158, in which
the brim is less contracted than the outlet. A line drawn
from the lower edge of the fourth lumbar vertebra, which had
fallen downward and forward to the symphysis pubis, measures
2½ inches, and there is a space across in an oblique line
of 3 inches. The outlet measurements are as follows: from
one pubis to the other below the symphysis, ¾ inch; from one
ischial tuberosity to the other, 1½ inch; from ischial spinous
process on right side to the upper part of the coccyx, ⅜ inch;
on the left side, ¼ inch; from one ischial spinous process to
the other, 1 inch.

The consultants in this case were hospital surgeons, and ranked high; a majority of them anti-Cæsareanists. Two or three proposed the Cæsarean section, but their proposition was overruled. The head of the infant was perforated, and small portions of the cranial bones were brought away, but the operator was unable farther to lessen its bulk and to draw it down.

She was abandoned to die with a mutilated infant having escaped into the abdomen. The perforator and common crotchet were the instruments used in this case. This case is a type of many others, and I think it indisputably proves that it is not as easy as it is represented to define the exact limits of pelvic brim contraction which are to stand as the immutable boundary which embraces the capabilities of craniotomy as an operation of election.

It has been stated that when the Cæsarean section is performed with the express object of saving both the life of the mother and the infant, "it is taking the woman's life in our hands, and deliberately subjecting it to the most imminent hazard for the sake of probably saving her child."

This is an assumption which neither morals, religion, nor obstetrical experience warrants.

My opinion on the comparative value of maternal and infantile life are already before the profession (see also Part I. page 63, which I most respectfully request the reader to refer to).

The late Sir J.Y. Simpson has stated as follows : "Assuredly no man would consider himself justified, on any plea whatever, in perforating and breaking down with a pointed iron instrument the skull of a living child an hour after birth, and subsequently scooping out its brain. But is the crime less when perpetrated an hour before birth ? Modern physiology has fully shown that there is no such distinction between the mental and physiological life of an infant an hour before labour is terminated and an hour after it as to make any adequate distinction between the enormity of the act, as perpetrated at the one or at the other of these two periods. And as if to add to the horrors of craniotomy when performed upon a living infant, some authors—and among them even the very latest—tell us, that whatever doubts may have existed as to the child

being alive or not at the date of operating, the results of the
operation itself will decide this point; for, if it be alive at the
time of the deadly perforation of its scalp, skull, and brain,
this fearful fact will be revealed to the practitioner by warm
and fluid streams of blood pouring along his fingers and hands
before any masses of broken brain escape, or the reverse."

Is it more true that "he who accelerates death is held
responsible for having caused death" in cases of great depressed
vitality in women from disease, in which cases Cæsarean section
is performed, than it is in cases of equal depressed vitality
from disease in which craniotomy is performed? ("Obstetric
Memoirs and Contributions," vol. i. p. 607.)   In the latter
case there is intentionally murder committed, which is trans-
lated into "justifiable homicide."

Dr. Barnes ("Lectures on Obstetrical Operations," page
421) dwells very much on the necessity of strictly adhering
to the moral law in reference to the performance of the
Cæsarean section; and I would ask whether it was not of
equal importance that the obstetrician should observe the
Mosaic law, "Thou shalt do no murder."

At page 422, Dr. Barnes asks, "Shall we dare to put a
mere vegetative life—that of an unborn child—into the scale
against that of a being like ourselves, accountable to the
Almighty?"

This low estimate of the life of an unborn child is remarkably at variance with social views.   Life is life, and the law
of the land and the conscience of the public equally recognize
its value.   How, then, can it be permitted to speak so lightly
of the life of a child as to say its life is vegetative?

In a course of lectures which I delivered to the recognized
members of the profession at the beginning of the year 1843,
when speaking of the comparative value of the life of the
infant *in utero*, I made use of a sentiment which I will tran-
scribe.   When the destruction of the infants by craniotomy is
contemplated, do we really consider the great social evil we
may commit by thus destroying an infant *in utero*?   We can
judge of the rotundity of the head, we can form an opinion
of its size and the degree of its ossification, and also of its
relative position, but we cannot judge of its future development?   There are no physical marks or phrenological indica-

tions from which we can ascertain by an examination *per vaginam* whether there exists in the brain such an organic condition as might enable the individual to become a most valuable and shining member of society.

I will just refer to one as an example. Suppose the head of Shakspeare had been opened, what would have been the loss to society ! Then, is the life of the infant of no value ? In the abstract, it may appear of comparative little value; but analyze the question in all its relative aspects, and the conclusion come to will be that great consideration is required. It is undoubtedly true that it is quite impossible to say what may be the result of the development of the brain of any child, and what may be the issue in after life.

In the present practice of obstetricy, the fullest consideration for the preservation of the life of the unborn infant has been extended to it by the use (to the most extreme limits) of the forceps and of turning, and in the powers and advisability of these operations I most cordially agree. But the life of the infant demands a more serious deliberation " even beyond the boundary above mentioned," which has been, and still is, exercised.

To prevent the reckless use of the perforator is an object of the highest importance, and claims the force of the restrictive influence of every obstetricist whose professional talents and reputation stand high in the profession.

The low estimate which is held of the value of the life of the unborn infant has led to a most unwarrantable abuse of the perforator and crotchet.

Dr. Ramsbotham says : " Of all instrumental operations in obstetric surgery, the perforation of the skull and extraction of the mutilated fœtus, is the easiest which could be undertaken for delivery in any case of impacted head, and much do I fear that to the facility with which this operation can be accomplished, have been sacrificed the lives of many children. It may, perhaps, be desirable in surgery, whenever necessity compels us to perform an apparently cruel operation, that the horror which the simple and bare mention of that act would inspire might be smothered and absorbed, as it were, by the sonorous and classical title which it bears ; but, by whatever name it is called, under whatever high-sounding appellation it is disguised, we cannot alter or conceal the fact that the operation consists in plunging an iron instrument into the centre of the skull of a human being, probably at that moment living, and extracting it after this mutilation has been practised."

This opinion I can fully verify. In the lectures I delivered to the profession in the year 1843, I mentioned cases which had been unwarrantably craniotomized. In one case the head was opened, and the child born alive; this doubtless awful catastrophe would accuse the operator of the great impropriety he had committed. In some cases the operation has been performed with such cold-blooded feeling as to use a penknife and domestic scissors.

The custom of the country has certainly justified craniotomy in reference to obstetrical practice, and there is no statement in the Legislative Code to warrant the performance of such an operation; but the question is, ought it to be recognized by the Legislature, and not stand, as it now does, as justifiable homicide, sanctioned alone by custom?

Dr. Simpson, the present Professor of Midwifery, Edinburgh, says : " He remembered, when taking his holiday in the English lakes, conversing with a doctor there, who related to him a case where he had to perforate, and where the only instruments he had were in his pocket-case. He had to perforate with a knife, and extract with a hook hastily manufactured by a smith who lived in the neighbourhood. The child was born alive, and survived."

There is another case related by Dr. Keiller, in which the practitioner used a pair of cobbler's pincers as a craniotomy instrument, which unquestionably crude mode of procedure, together with its unfortunate results, led not only to legal investigation, &c. (See *Edinburgh Medical Journal*, May, 1880, pp. 1032 and 1033.) The above statements warrant me to declare that craniotomy, whether it stand as an operation of election or as one of necessity, ought never to be performed without its propriety has been decided by a consultation. The necessity of such a step ought to be enforced by writers and lecturers on midwifery.

I have been asked by an eminent obstetricist if I dispute that craniotomy may not be resorted to " to save the mother's life." In answer to this question, I beg most unequivocally to say, " Yes," when her life is jeopardized by some temporary existing cause; and under other circumstances, I should conditionally make craniotomy no operation of election; but neither my judgment nor my conscience would ever lead me to such a conclusion, because I considered the Cæsarean section a more hazardous operation than craniotomy in cases of extreme deformity of the pelvis.

Two cases, in one of which there is only a limited degree of contraction of the pelvic brim, and in the other an extreme degree of diminution in this aperture are not analogous.

In one, "the first," craniotomy will only be required to be performed ONCE, as, in the event of subsequent pregnancies occurring, *premature labour* may be induced, and *the infant be thereby saved;* whereas, in the other case, craniotomy would have to be repeated time after time. In such cases, my conscience tells me such horrible destruction of life cannot be defended on any ground whatever. (See Part I. p. 64.)

As I have before stated, it is quite impossible to ascertain with such mathematical accuracy the precise measurement of an extremely contracted pelvic brim, in which a mistake of a very small fractional diminution would render the extraction of the base of a full-sized infant's head quite impossible.

If this opinion be true, Cæsarean section and craniotomy are not equally influenced by it. If craniotomy is performed under a miscalculation of the exact pelvic measurements, and delivery *per vias naturales* cannot be effected, then, after sacrificing the infant, the woman must be left to perish, or she must undergo the Cæsarean section; whereas, if the Cæsarean section had been performed as an operation of election, the lives of both mother and infant might have been preserved.

The hands of different individuals differ very considerably in size, and therefore a computation made on such a basis, in order to use the cephalotribe, must be liable to lead to an erroneous conclusion. It may be conceded to eminent obstetricians in pelvic explanatory inquiry what must be denied to the great majority of practitioners, and as I have already stated the measurement of the brim may be more than is said to be required for craniotomy, and yet delivery cannot be effected.

No doubt the success of both these operations would be very different if they were duly performed before irretrievable mischief was inflicted by protraction. We have no statistics of craniotomy when performed under high pelvic distortion ; but we have those (which I have elsewhere given) of Cæsarean section, which show great maternal mortality, and which consequently weigh against this operation ; but if an unprejudiced analysis is made of them a different conclusion must be come to. They very decidedly show the mischievous results of protracted labour and other factors.

It has been asserted by the late Dr. D. Davis, and echoed by a later writer, that recourse to the Cæsarean section, " is justly called" the last extremity of our art and the " forlorn hope of the patient ;" and I think it may be said with as much if not more truth and propriety, that there is no operation more revolting to human nature than craniotomy, which is, in fact, direct murder of a human being, and may be with great truth pronounced as a dreadful expedient.

See Table II.

Statistics.—The results of the Cæsarean section in reference to the mothers are very unfavourable ; of the 19 women whose cases are tabulated, 15 died, or 78·94 per cent. ; four, or 21·05 per cent., were saved. In one of these cases, the report extends only to 48 hours after the operation. Further information I have not received, although I sought for it. From the 19 women the same number of infants were extracted, of which number, 13, or 72·22 per cent. were saved.[*] Two of this number afterwards died, one being very small and feeble ; the other premature, being only six months and a-half. These two infants were the offspring of women suffering from cancerous degeneration of the cervix and os uteri. Of the entire number, five, or 26·31 per cent., were dead when extracted.

These two cases (one saved, and one lost), added to the nineteen which were tabulated in my last communication, make twenty-one cases. The one (97) added to the four before reported gives five, or 23·80 per cent. saved. The other (98) case added to the number (fifteen) last registered, gives

---

[*] In one case (82), there is no account whether the infant was living or dead ; if living, which is most likely, the number saved would stand 14, or 73·68 per cent.

sixteen, or 79·01 per cent., lost. The two (infants one dead, one alive), added to the nineteen already mentioned, make twenty-one, of which fifteen, or 71·42 per cent. were preserved, and six, or 28·57 per cent., were lost. The one dead, belonging to the woman (Case 97) was dead when forced out from the uterus, and was non-viable, being only at five months. (See page 11.)

SEE TABLE II.

Maternal Mortality.—The maternal mortality is shown above to be very great; and it is of the highest importance to ascertain what has caused such a fatality. We find that the constitutional state of nearly all the women who underwent this operation was in a most unfavourable condition to bear it without great hazard; and in several of them the pelvic organs and tissues had suffered from pressure. Seven of these women laboured under progressive and incurable disease, with four of whom it was malignant, there being cancerous degeneration of the cervix and os uteri, and one of the rectum; one was nearly bedridden, and suffered from mollities ossium. Four women were rickety, and greatly exhausted; one of them was a dwarf, and had had craniotomy unsuccessfully performed after a very long labour. In one the infant presented footling, and, after fruitlessly drawing down the feet, breech, and arms, the head was perforated, and the cephalotribe unsuccessfully attempted to be used. The body was afterwards dragged away, leaving the head behind, which afterwards escaped into the abdomen through a rent of uterine tissue. One was greatly exhausted, and suffered from considerable pelvic mischief. Another suffered from great exhaustion, and had endured a long labour; the infant's arms presented; turning was attempted, and the blunt hook and cephalotribe were unsuccessfully used; the muscular tissue of the uterus was torn. Two of the women who died were not in a bad constitutional state. One of them had a large fibrous tumour, which blocked up the pelvis. In the other, pelvic deformity existed, and she had labour nduced at eight months and a-half. In two cases we have no account.

The woman (Case 98) was rickety. She died four days after the operation.

TABLE III.

Of the 33 cases recorded in this Table, 26 died, showing a mortality of 78·78 per cent.; the causes of death varied and are mentioned in the respective sections; there are, however, four cases which require special attention—viz., Nos. 118, 125, 126, and 130—in each of these craniotomy had been performed, and it was not until after prolonged efforts had failed in delivering the women *per vias naturales* that Cæsarean section was resorted to.

One case, No. 106, was thought to have ended fatally through an embolism in the heart. Two, Nos. 112 and 120, were due to septicœmia, and one, No. 127, which took place on the fourteenth day, and after the wound had completely healed, was caused by a severe attack of diarrhœa; the gentleman who reported the case did not think that there was any connection between the diarrhœa and the operation. One death, No. 107, is not accounted for. The remaining deaths, 17 in number, are inserted according to the cause in the respective sections. (See page 13.)

The following is a statement of the number of deaths which have occurred, and are classed under the different causes which are recorded in the three Tables, as rendering the Cæsarean section necessary :—

TABLE I.

In cases of mollities ossium, 41; in cases of rickets, 12; in cases of exostosis, 2; in cases of tumours, 5; in a case of fracture of the pelvis, 1; in four cases no account of cause is given.

TABLE II.

In cases of mollities ossium, 2; in cases of rickets, 5; in a case of exostosis, 1; in a case of tumour, 1; in a case of exostosis, 1; in cases of cancer of os and cervix uteri, 2; in a case of cancer of the rectum, 1; in a case of oblique distortion of the pelvis, 1; hydatid cyst, 1; in one case the cause is not stated.

TABLE III.

In cases of mollities ossium, 8 ; in cases of rickets, 9 ; in a case of exostosis, 1 ; in cases of cancer, 6 ; in cases of tumours, 2 ; in one case the cause is not stated.

Total number of deaths in cases of mollities ossium .			. .	51
„	„	„	rickets	26
„	„	„	exostosis	4
„	„	„	tumour	8
„	„	„	cancer of os and cervix uteri	8
„	„	„	oblique distortion of the pelvis	1
„	„	„	hydatid cyst	1
„	„	„	of fracture of pelvis .	1
„	„	„	in cancer of rectum .	1
„	„	„	not stated	7

108

108, or 82·44 per cent.

SEE TABLE II.

Infantile Mortality.—Infants *in utero* may die during pregnancy from disease within their own bodies, or from any morbid cause existing in the placenta or funis which can interrupt the supply of blood from the maternal system. Undue protraction of labour is extremely hazardous to the lives of infants, and especially so after the discharge of the liquor amnii. The five infants which are stated to have been dead were so before the performance of the operation. Of this number, two were doubtless destroyed by long-continued pressure during protracted labour. One presented with the arm, for the delivery of which manual and instrumental attempts had been unsuccessfully made. One presented with the feet, in which case, after drawing down the body, craniotomy was performed; and afterwards the body was dragged away by the feet, leaving the perforated head behind. In another, craniotomy was unsuccessfully performed. It is quite evident from the above statement that there is not a single infantile death which was caused, either directly or indirectly, by the operation; and there is little doubt that all of them would have been saved, if they had been living at the time of its performance.

Two (infants one dead, one alive) added to the nineteen already mentioned, make twenty-one, of which fifteen, or 71·42 were preserved, and six, or 28·57 per cent. were lost. The one dead, belonging to the woman (Case 97) was dead when forced out from the uterus, and was non-viable, being only at five months.

TABLE III.

The number of infants reported as being dead was eleven, which is a mortality of 33·33 per cent.

This death-rate does not truly represent the infantile mortality after Cæsarean section, for four of the above-named infants were killed by craniotomy—viz., Nos. 118, 125, 126, and 130; and four—viz., Nos. 111, 113, 119 and 122—were dead prior to the commencement of the operation, so that in reality there are not more than three deaths, or 9·09 per cent., in any way chargeable to the operation. (See page 24.)

INFANTILE DEATHS.

General Statement of Infantile Deaths, as follows :—

No. of Case in the Table.	Duration of Labour.	No. of Case in the Table.	Duration of Labour.	No. of Case in the Table.	Duration of Labour.
1.	12 days	27.	37 hours	60.	12 hours
2.	7 ,,	30.	22 ,,	61.	70 ,,
5.	no account	33.	30 ,,	65.	30 ,,
6.	,,	34.	20 ,,	66.	3 days
7.	30 hours	39.	3 days	67.	18 hours
10.	no account	40.	10 hours	69.	4 days
12.	5 days	41.	53 ,,	72.	4½ ,,
14.	54 hours	43.	30 ,,	73.	17 hours
15.	10 days	44.	no account	75.	18 ,,
16.	3 ,,	46.	40 hours	79.	8 days
19.	81 hours	47.	10 days	80.	72 hours
22.	3 days.	51.	3½ ,,	83.	96 ,,
24.	40 hours	56.	84 hours	84.	28 ,,

No. of Case in the Table.	Duration of Labour.	No. of Case in the Table.	Duration of Labour.	No. of Case in the Table.	Duration of Labour.
85.	112 hours	111.	2 days	122.	6 to 7 days
102.	24 ,,	113.	28 hours	125.	32 hours
109.	2 days	114.	a long time	126.	no account
110.	23 hours	118.	6 to 7 days	129.	31 hours
		119.	early	130.	no account

Total   53

Embryotomy   1

Craniotomy   9

Dead before the operation   16—26

27, or 50·94 per cent.

The aforesaid statements demonstratively show that the deaths of the infants caused by the operation are comparatively few, being only 27; but I believe that if the cases had been drawn up with the care they ought to have been, many more infants would be found dead before the operation.

INFANTS SAVED.

General statement of infants saved, 78, or 59·54 per cent.

SEE TABLE II.

Exhaustion.—Exhaustion is stated to be the only cause of death in eight cases; in one of which labour is stated to have existed some days—eight days after the rupture of the membranes. Several of the abdominal organs were diseased; there was also a large hydatid cyst, extending from the liver into the pelvis, which was the obstructive cause of labour. In one the membranes had ruptured ten days before, and active labour had existed several hours. Several of the abdominal organs were diseased, and the os and cervix uteri were affected with cancerous degeneration. One woman, who had not commenced labour, sank on the sixth day after the operation from the exhaustive influence of a large cancerous growth. One woman was four days in labour, and was in a bad state of health when it commenced. In one the duration of labour was, as far as can be computed, from seventy to eighty hours; the membranes had ruptured ninety-six hours before the opera-

tion was performed, and several unsuccessful manual and instrumental attempts had been made to deliver the woman;
rupture of the uterus had taken place. Another patient was
only in labour twenty-eight hours, or thereabouts, after the
discharge of the liquor amnii, but craniotomy had been unsuccessfully performed, and the infant's body had been forcibly dragged
away, leaving the mutilated head *in utero*, which afterwards
passed into the abdomen through a rent in this organ. One
woman, a dwarf, is reported to have died exhausted, who was
in labour seventy-two hours; her infant had been unsuccessfully
craniotomized. The death of one woman is assigned to exhaustion; she was in labour only twelve hours, and delivery
was obstructed by a large fibrous tumour. On an attentive
consideration of the circumstances of the aforesaid cases, there
exists abundant evidence in all (with the exception of the last
case) of what were the real causes of the failure of the vital
powers.

TABLE III.

In twelve cases death was owing to exhaustion; in five of
these the women suffered from violent and extensive carcinomatous disease of the uterus, or vagina; in one of these,
No. 120, the vagina was so blocked up with a malignant
growth as not to permit of the passage of a No. 7 catheter
into the uterus, and she was not admitted into the hospital
until the seventh day of labour, and she was so reduced by
exhaustion and irritative fever that she died before the operation could be completed. In two cases, Nos. 118 and 125,
death was caused, and was entirely due to the prolonged and
fruitless attempts at delivery after the performance of craniotomy. In five cases labour was allowed to be so long protracted as to reduce the vital powers to so low a degree as to
be apparently the main cause of death from the great exhaustion
produced. (See page 16).

SEE TABLE II.

Peritonitis.—Peritonitis is recorded as the cause of death in
six cases. In one, the duration of labour was sixty-six hours;
the woman was likewise in a bad constitutional condition. In

one it was 112 hours. This woman was also in a bad state of health. In one case, the woman was in labour forty-eight hours; and at least thirty to forty hours after the discharge of the liquor amnii. The cervix uteri was soft and dark coloured, and the os was patulous and nearly black, as if gangrenous. The internal surface of the uterus was darker and rougher than natural. In one case, labour had lasted ninety-six hours; and the woman was greatly exhausted at its commencement. In one case, it is stated, the waters had been discharged ten days before the operation; but active labour had only existed three or four hours. She had cancerous degeneration of the cervix uteri, and was completely anæmic from previous losses of blood, and there existed disease in the liver, spleen, and other organs. In one case there was "slight injection of the peritoneum." The operation was performed at eight and a half months of pregnancy. Labour was induced by "secale." Long duration of labour stands prominently forward in nearly all those cases of Cæsarean section in which death has been recorded as having been caused by either *exhaustion* or by *peritonitis*. Protraction in labour ought at all times to be considered as a factor of danger. The mischief inflicted varies in degree in different cases, according as there may exist different relative circumstances; but it is an indisputable fact, that danger nearly always increases in proportion to the duration of labour. There are doubtless other contingent conditions which aggravate the effects thus produced; such as the early evacuation of the liquor amnii; the kind and degree of the obstacle which impedes the progress of the infant through the pelvis; or a low constitutional and local state of vital power. In most of the aforesaid fatal cases, some, if not all, of these elements of mischief existed.

The cause of death in Case 98 was peritonitis. She appeared to progress favourably until the third day, when the disease came on. But it is quite evident that the peritoneum could not have been in a healthy condition previously to the operation; for, as soon as the incision was made through the abdominal parietes, some ounces of serum escaped. The low vital state of this membrane must most assuredly have been greatly damaged by the after-treatment. Dr. Hardin says:

" Her chances of recovery were considerably lessened by want of proper care and nourishment." The woman (Case 97) had a pretty smart attack of peritonitis. It commenced on the second day, and became worse on the third. It was, however, subdued and auspiciously terminated.

TABLE III.

Nine deaths are recorded which were due to peritonitis, one, No. 99, who was suffering from a large fibroid tumour attached to the sacrum, and which filled the inlet of the pelvis. One, No. 103, appears to have been for the most part due to the labour having been too long protracted. Two were distinctly of a septicæmic origin, namely, Nos. 116 and 117. Two, Nos. 126 and 130, by the operation of craniotomy which preceded the Cæsarean section. Another, No. 118, was due to carcinoma ; and the remaining two may ostensibly be attributed to the mischief caused by protracted labour. (See page 19.)

SEE TABLE II.

Hæmorrhage.—Hæmorrhage during labour is an accident which is generally attended with great peril to the mother, and claims the gravest consideration and attention of the obstetrician. Every case, however slight it may appear, claims his anxiety and care. There is, however, one consolation : he has the means, if he exercise judgment, to arrest the danger of some of the most appalling cases. But it may be reasonably supposed that hæmorrhage is a more serious factor of danger in Cæsarean cases than in labours in general. The sources whence blood is discharged, during and after the operation, are from the incised edges of the abdominal and uterine wounds; and when the placenta is located under the line of the incision, it may be cut, and the blood issues from its divided structure ; or, if this organ be torn, it then flows from its disruptured tissue. So, also, if the placenta be detached from the uterus, when atonic, then blood is poured out from the sinuous openings. It is a remarkable fact that there has been very little blood lost in the great majority of the nineteen recorded cases of Cæsarean section. In eight of these cases the placenta adhered to the anterior part of the uterus, in four of which there was very little bleeding. The placenta was not cut during the operation, but it was very cautiously and only partially detached, so as just to admit the hand of the operator. But in the other four of these cases (placenta anteriorly placed) the bleeding which proceeded from the edges of the uterine wound was profuse; in two of them the placenta was cut into, and in the other two cases this organ was completely separated and removed by the operator before he passed his hand into the uterus to extract the infant. In eight of the entire number the placenta was situated on the posterior part of the uterus; in two of the cases there was a considerable loss of blood; in five there was very little lost; and in one there was no bleeding. Chloroform was administered in ten cases; in five of which there was only a little blood lost; in four the bleeding was very great. Ether-spray was used in four cases, in which there was very little blood lost. The factors of hæmorrhage in Cæsarean cases are the same as those which operate in all kinds of labour, with an additional one, the incised edges of the abdominal and uterine wounds. Whatever diminishes the contractile power of the uterus nearly always produces this accident. It has already been mentioned that there is no cause which is more influential in weakening the vital energy of the uterus than protracted labour. We find that those who lost the most blood were in labour for a considerable length of time. One was from seventy to eighty hours; one 112 hours; one sixty-six hours; and in one the membranes had been ruptured ten days, and active labour had existed several hours. This woman suffered under cancerous degeneration of the cervix uteri, and her general health was very much impaired by loss of blood and other discharges which she had during her pregnancy.

Hæmorrhage happened in one case, No. 97. There was, as already mentioned, bleeding from three arteries at the lower part of the lacerated tissue, which was effectually subdued by silk ligatures. Oozing of blood from the placental surface of the uterus continued; but it was checked by the introduction of a piece of ice into the cavity of the uterus, and also by firmly grasping this organ. Mr. Wells passed his finger *per vaginam* and os uteri, in order to make a free passage for the discharge, and thereby prevent its passing into the abdomen. Two hours after the operation bleeding became rather free, but it was at once checked by a drachm dose of the liquid extract of ergot.

TABLE III.

Hæmorrhage occurred in twenty cases of those recorded. In eight only was it sufficiently profuse to place the patient's life in immediate danger. Of these eight cases there were three—viz., Nos. 103, 107, and 121—in which the source of the hæmorrhage was both from the uterine incision and the placental site. In Nos. 106, 109, 110, and 119, the hæmorrhage came almost entirely from the uterine incision, while in No. 112 the hæmorrhage was solely placental.

In the remaining twelve cases the hæmorrhage was not so profuse, and the sources from whence it came varied; in four the hæmorrhage proceeded both from the uterine incision and the placental site; in three cases it was almost altogether from the placental site alone; and in five cases the source of hæmorrhage was for the most part from the uterine incision. (See page 17.)

TABLE II.

Successful Cases.—The tabulated record contains four cases of recovery after the Cæsarean section. In one case, enumerated as one of success, I can only state what has been given to the profession. It is registered No. 91 in the Tables; and the report made by the operator only extends to forty-eight hours after the operation, when it is stated that she was doing well. Further information I have been unable to obtain. In Case 89, the cause of impediment to the passage of the infant through the pelvis was epithelioma of the cervix uteri. Dr. Greenhalgh operated before labour had commenced, and before the membranes had ruptured; and doubtless we are indebted to this gentleman for its success by adopting an early operation, and for its judicious treatment afterwards. The case merits attentive perusal and deep consideration. Ether-spray was used. In Case 88, the impediment to labour arose, again, from extensive epithelioma of the cervix and lower part of the body of the uterus. The labour in this case had commenced prematurely at six months and a half. Ether-spray was used.   Mr. Newman's judicious and firm treatment of the case merits the approbation of the profession. The recoveries of these two women, labouring under malignant disease, and having undergone such serious operations, are truly wonderful, and afford the strongest evidence of the conservative powers of Nature.   These cases ought to teach us to have more confidence as to the result of this operation ; and they afford contradictory proof against the assertion of those practitioners who despair of success, and look upon the reductive changes of the puerperal uterus as antagonistic to recovery. The other case of recovery, No. 96, is most interesting and satisfactory.   The pelvic obstruction was caused by exostosis springing from the sacrum. (See Table.)   The patient was in good health ; and the operation was early and judiciously performed by Mr. I. B. Brown, with the concurrence and assistance of Dr. John Taylor, whose patient she was, before the constitutional and local vital powers had suffered from pressure, &c. The membranes had not ruptured. There are many important practical points which Dr. John Taylor communicated to me, some of which have already been used, as the location of the placenta, and the hæmorrhage which issued from the uterine veins and walls.   There are other matters which shall be spoken of under their respective heads.

The recovery of the woman, Case 97, from such a complication and number of very serious matters affords further and still stronger evidence of the wonderful conservative powers of Nature.   A careful perusal of the case, as detailed by Mr. Wells, ought to convince the most violent and prejudiced anti-Cæsareanist that this operation ought to be recognized and performed with the same confidence of success as that with which other capital operations are now undertaken.

No. of Case.	Duration of Labour.	Cause of Difficulty.	No. of Case.	Duration of Labour.	Cause of Difficulty.
1	12 hours.	Not stated.	88	4 days; liquor amnii abstracted 2 days before the operation.	Epithelioma of the cervix and lower part of the body of uterus.
12	5 days.	Distorted pelvis, from fracture.			
35	About 34 hours.	Mollities ossium.	89	Labour had not begun; membranes not ruptured.	Epithelioma of the cervix uteri.
36	About 30 hours.	Mollities ossium.			
37	24 hours.	Mollities ossium.	91	Not stated.	High distortion of pelvis; conjugate diameter, 1½ inch.
49	12 to 14 hours.	Mollities ossium.			
53	Slight pains for 2 to 3 hours; membranes unruptured until a few hours before the operation.	Mollities ossium.	96	18 hours; membranes not ruptured.	Exostosis.
			97	Labour did not exist.	No pelvic distortion; ovarian disease. A mistake.
57	Indefinite slight pains for 2 or 3 days; membranes ruptured 12 hours before operation.	Cancer of the cervix uteri.	104	2 days.	Mollities ossium.
			110	23 hours.	Rickets.
			111	Slight pains for 2 days.	Rickets.
			119	Early.	Rickets.
67	18 hours.	Exostosis.	121	24 hours, but not severe labour.	Rickets.
68	Not stated.	Rickets most probably.			
71	6 days.	Hard cancer of os and cervix uteri.	123	60 hours.	Exostosis.
			124	35 hours.	Rickets.

SEE TABLE II.

Chloroform.—Chloroform is nearly always administered in capital or important operations, for the purposes of inducing a state of quietude in the patient, of diminishing the pain inflicted by the knife, and of lessening nervous shock. To attain these ends it is most important and valuable. In most operations there are few objections to be raised against its administration; but in Cæsarean cases it is otherwise. For the success of the operation it is essential to obtain a full, complete, and energetic contraction of the uterus. This organic condition secures against hæmorrhage, and, both immediately and remotely, not only lessens the size of the wound, but also contributes to a nice and firm adaptation of its edges. Therefore every method should be adopted to aid this organic state; and everything ought to be avoided which interferes with it or lessens it in the least degree. Chloroform diminishes the energy of the uterus, and induces relaxation; and, therefore, on this ground, its administration is here contra-indicated. Its anæsthetic agency in subduing pain is not to be compared to the dangers which may ensue. I have never known the moral courage of women fail, and have always found that they have endured the operation with great fortitude. A common expression has been, that they have suffered less pain during the operation than they have had from one unavailing labour-pain. Vomiting has taken place in most of the cases in which chloroform had been administered. It has varied in degree. In some cases it has been very urgent and continued. In Case 96 it was incessant until the fourth day; the abdominal wound was burst open; a knuckle of intestine was protruded; and fresh sutures were required. Chloroform was inhaled in ten cases; vomiting followed in nine. Whether this effect is to be considered as a *propter hoc* or a *post hoc* some may doubt; but I am disposed to consider it as the first. At all events, the effect of vomiting after this operation is most hazardous. Dr. Kidd endeavours to trace vomiting after chloroform to other causes. (See his remark, *Edin. Med. Jour.*, January, 1868, p. 596.)

Chloroform was administered in both the cases. For my opinion upon the influence of this anæsthetic on the uterus, the reader is referred to my two former communications. Vomiting occurred afterwards in both cases. On the question whether the vomiting was a *post hoc* or a *propter hoc*, I again refer my readers to observations already made. Mr. Wells, in his letter to me, states: "I don't think we can judge very well what share chloroform had in the sickness." The incision made in the abdominal parietes was in both cases longitudinal and along the linea alba.

TABLE III.

Chloroform was administered in most of the cases; in one, No. 117, methylene was used; in another, No. 122, ether was employed: in two cases ether was substituted for the chloroform, and in one case, No. 111, no anæsthetic was given, and this patient had no vomiting at all, and but little hæmorrhage occurred during the operation.

Vomiting frequently occurred after the operation, and seemed in some instances to increase the exhaustion very much. How far the vomiting could be attributed to the anæsthetic it is difficult to say, especially as no mention is made as to the presence of this dangerous complication.   It is, however, worthy of remark, that in the case in which no anæsthetic was employed, no vomiting took place.

Dr. Barnes says : " Dr. Keith, whose success in ovariotomy is so conspicuous, has entered a decided protest against this operation.   ' Had chloroform,' he says, ' never been heard of, I doubt if humanity would have suffered from the want of it.' "   (See page 29.)

SEE TABLE II.

Ether-spray.—The profession is greatly indebted to Dr. Richardson for the introduction of ether-spray as a local anæsthetic.   This mode of subduing pain inflicted by the knife has been employed in four Cæsarean cases, and doubtless it is a most valuable agent.   It has been remarked that it not only completely subdues sensibility, but it also promotes energetic contraction of the uterus.   In all the four cases the abdominal parietes were completely æsthetized ; but in two only, as far as I can ascertain, was the ether-spray played upon the uterus.   In both these cases strong uterine action was induced, which seized and firmly embraced the bodies of the infants, and thereby rendered their extraction difficult.   In one of these cases, " the uterus contracted so firmly as for a short time to impede the introduction of the hand into its cavity."   Uterine contraction after the removal of the infant and the placenta is the most effective security against the dangers of hæmorrhage, &c. ; but if it be inordinate in degree and prematurely induced, it then becomes mischievous and dangerous to the life of the infant, by firmly grasping it, and rendering its withdrawal extremely difficult.   There is also some risk of the uterine tissue being lacerated, as the length of the wound becomes comparatively so much diminished by the contraction of the womb.   In one of the cases, No. 93, at which I was present, the uterine tissue appeared to be har-

dened, and thereby rendered less fit for a nice adaptation of
the edges of the wound. They presented a gaping character,
which was very conspicuous after death. Serious reflection on
the effects of ether-spray which I witnessed in this case, No. 93,
and also on those reported to have happened in the other cases,
has convinced me that the application of the anæsthetic agent
should be solely confined in Cæsarean cases to the abdominal
parietes.   I agree with Dr. Roberts that the uterus should not
be subjected to its influence, because I feel certain that there
is less mischief to be apprehended from the pain inflicted by
incising the organ, than what is likely to accrue from the
playing of the ether-spray on the part when the abdomen is
laid open. I do not apprehend that the edges of the abdominal
wound are likely to suffer in the way Dr. Kidd (*loc. cit.*) states.
He says, " Under ether-spray the lips of the wound, too, may
mortify."

See Table II.

Sutures.—The edges of the uterine wound have been
brought together by sutures in three of the cases. In two
there was considerable hæmorrhage ; and in them this practice was adopted to restrain the bleeding, and to bring together
the edges, which were gaping. In one, iron wire was used ;
in the other, common ligature. In the third, the interrupted
suture was employed. These women died. Although I am
not disposed to attribute their deaths to the application of the
ligatures, yet I am strongly of opinion that sutures ought not to
be introduced into the uterine walls ; and, indeed, I think that
they would prove, in general, not only useless, but, by the
uterine tissue yielding, they would be injurious. Since the
above was written, a case of recovery after Cæsarean section, No. 96, has occurred, in which the edges of the uterine
wound were brought together by eight silver sutures, which
were left in the uterus. The abdominal wound was also closed
by silver sutures.

Both abdominal and uterine sutures were used in each of
the cases. In Case 97, Mr. Wells brought the edges of
the uterine wound together by an uninterrupted suture of
fine silk, one long end of which was passed into the uterine

cavity, and out through the os uteri into the vagina. The edges of the uterine wound were accurately adapted by seven or eight points. The other end of the silk was brought through the abdominal walls along with the three ligatures which were applied to the bleeding arteries : and all the ligatures were tied to the clamp, which embraced the ovarian pedicle. It is worthy of remark, that Mr. Wells burnt off by a hot iron a portion of the omentum; and, as there was free bleeding, he afterwards applied three silk ligatures; and, after cutting their ends short, he returned the omentum into the abdomen. The edges of the abdominal wound were adjusted by one superficial and six deep silk sutures. In the other case, No. 98, as " the uterine incision was somewhat serrated, it was deemed advisable to put in a couple of metallic sutures. Several others were put in the abdominal walls." In a former publication I stated that I had an objection to the uterine sutures, as I considered them likely to be injurious. I then mentioned one case of recovery that had taken place in which metallic sutures were applied. We have here another example of recovery. Mr. Wells used the uninterrupted silk suture. He thought that the escape of blood or of the secretion from the uterine cavity of the peritoneal cavity might be one cause of the mortality of the Cæsarean section; and, if so, sutures might be useful. If the reader refer to my former observations (*British Medical Journal*, April 4, p. 321), he will find that I have already adverted to this danger, and referred to a case, No. 95, in which the death from peritonitis was attributed to the passage of the uterine discharges into the peritoneal cavity, which were prevented from flowing *per vaginam* by the closed condition of the os and cervix uteri. Whether uterine sutures should be employed, is a most important question to decide. There are cases of recovery after this operation in which sutures were not used, as well as cases of recovery in which they have been applied. But, as regards the evidence to be obtained on the two sides of the question, it would, I think, preponderate in favour of their non-employment. Notwithstanding this may be so, yet I think the subject demands further consideration. If I used a suture, I should prefer the uninterrupted, put in at long points, and with only one long end passed through the os uteri into the vagina; the other end closely cut off, according to the last opinion of Mr. Wells.

TABLE III.

In sixteen cases the edges of the uterine wound were brought together by sutures; in seven cases carbolized catgut was the material used; in three cases, wire sutures were employed; in four cases, silk sutures were used; in one, fish-gut; and there were two cases where the kind of suture used was not known.

It was noticed at some of the post-mortem examinations that the uterine wound was gaping. This unfavourable condition was chiefly noticed in those cases in which carbolized gut was used.

The abdominal incision in each case was closed by sutures; in one, the quill suture was used; in eight cases, wire was used; in thirteen, silk; and in twenty cases, the kind of suture is not mentioned.

SEE TABLE II.

Operation.—I shall not here enter into all the details which are requisite to be observed before, during, and after this operation; but I shall confine my remarks to some of the more important points. (See observations, *British Medical Journal*, vol. i. 1865, p. 263.)

Frequent examinations *per vaginam* are extremely injurious, from the contusions which the vagina and pelvic tissues sustain. There are cases recorded which forcibly prove the necessity of this precaution. When the uterus is deflected, as it generally is in cases in which there is extreme distortion of the pelvis from mollities ossium, this organ must be raised up before the incision is made, so that the fundus, which abounds with large anastomosing veins, may not be incised. Neglect of this rule doubtless would produce hæmorrhage by the division of this portion of the uterus, which is so vascular and eminently contractile, and consequently would interrupt the efficient contraction of the organ, which is so important to produce a nice adaptation of the edges of the wound, and thereby, in some measure at least, to prevent it from assuming the gaping character which occurred in some of the tabulated cases.

I

Labour unduly protracted, from any cause whatever, is nearly always attended or followed by some danger to either the mother or to her infant, or to both. The degree of mischief inflicted on the maternal structures is in general in a ratio proportioned to the period of protraction ; and, doubtless, in all cases it is aggravated by the nature of the obstructing cause. We ought, therefore, to be extremely watchful in all cases of protracted labour, but especially so when the impediment arises from a mechanical cause ; and we should timely adopt appropriate measures for the delivery of the woman before irreparable mischief is done. (See former observations, *British Medical Journal ;* also Cases of Laceration of the Uterus, &c., *Obstetrical Transactions*, London, vol. viii. p. 210.)

In ordinary labours, where the infant can pass *per vias naturales*, certain organic changes are waited for ; the os uteri must be more or less dilated ; but in those cases which demand the Cæsarean section, it is not only a very great folly to wait, as dilatation cannot be effected, but it is a very great evil.

An early performance of the operation is of the utmost importance. Supposing the woman at full term of pregnancy, it should be commenced as soon as the labour is declared— before, or at least immediately after, the membranes are ruptured. By so doing, the uterine incision would be relatively considerably diminished after the complete contraction of the uterus, and all the dangers of protraction avoided. My friend Dr. Greenhalgh has suggested the performance of this operation before the completion of pregnancy. He says : " I have a strong conviction that greater success would attend our endeavours if all cases were operated upon at, or shortly after the completion of, the eighth month of utero-gestation, when the vessels are smaller, the contractile power of the uterus greater, and the liquor amnii relatively larger in proportion to the size of the child. By pursuing such a course, a smaller incision would be required, less blood would be lost." I am inclined to agree with the above suggestion, especially if the obstructing cause of labour be epithelioma of the os or cervix uteri. Two successful cases are recorded in which the operation was performed before the completion of pregnancy— one at the end of the eighth month, by Dr. Greenhalgh ; the other at six and a half to seven months, by Dr. Newman. There is, however, one objection, either real or imaginary, against having recourse to the operation before the completion of pregnancy, when the os uteri is closed by the mucous plug, and the cervix is undeveloped; which condition would tend to prevent the lochial or other fluids from issuing from the cavity of the uterus. In consequence of this opposition, the fluid would accumulate up to a certain amount, and then be discharged into the abdominal cavity. Such an event happened (closing of the os and cervix uteri) on the second day after the operation in Case 95. Care must, however, be taken to keep open, if possible, the oral and cervical portion of the organ by some means or other.

Before the incision is made, it is of great importance to ascertain the location of the placenta. Its position in sixteen cases was as follows:—In eight it was placed anteriorly; in seven of which it was central, in one a little to the left. In eight cases it was fixed on the posterior part of the uterus; in three of which it was centrally situated; in two, it was towards the fundus; in one, to the left; and in two, to the right side.

During the operations, the placenta was unfortunately cut in two cases; it was partially separated in four cases; and in two cases it was completely detached and removed before the extraction of the infant, which is most hazardous to the mother (see remarks on Hæmorrhage), and also to the infant. In order to avoid as far as possible cutting the placenta, the stethoscope ought to be used; and doubtless, in the great majority of cases, satisfactory information, either negative or positive, will be obtained. If the "placental soufflet" is not heard, the infant being still alive, it is fair to conclude that this organ is not within the reach of the knife. If the infant is dead, there is not much risk of cutting into the placenta.

In sixteen cases the incision was made longitudinally in the centre along the linea alba; in three cases it was made a little to the left of it. I prefer the left side, at a little distance from the linea alba.

It is of the greatest importance that the operation should be completed as expeditiously as possible, in order to avoid the hazard of inducing rapid and irregular contraction of the uterus before the incision is made sufficiently long, so as to safely extract the infant. It is of great consequence not to have again to incise the organ when the infant's body is in a great measure brought out. This has happened. In Case 96 the uterine tissue was torn, in consequence of the incision being too small; it was only four inches long.

Few remarks need be made on this part of the subject. The location of the placenta should, if possible, be ascertained by the stethoscope. In fact, this instrument ought to be carefully employed in all cases in which the abdominal cavity is intended to be laid open, either for the removal of an infant from the uterus, or for the extirpation of an enlarged ovarium. The necessity of observing this rule is most forcibly proved by the results of Case 97. Mr. Wells has kindly informed me that the placenta in his case was situated anteriorly and towards the left side. In the other case, No. 98, the seat of the placenta is not mentioned; but it is presumed that it was on the posterior part of the uterus. (See page 24.)

I will now quote a few remarks on this operation from Dr. Harris.

GREAT as are the admitted dangers of craniotomy, cephalotripsy, and embryulcia to the mother, there are those who appear to hold to the opinion that we should never make choice of the Cæsarean operation, if the fœtus can by any possibility be delivered *per vias naturales*. We have had women in the United States who endured several hours of suffering under craniotomy, and narrowly escaped with their lives, who were afterwards delivered safely of living children by gastro-hysterotomy. In fifteen of the 100 cases reported, the operation was predetermined on account of former, or anticipated, difficulties, and the same arranged for by the operator. In thirteen the women recovered, and all of the children were delivered alive but one, the child presenting by the arm. In five instances the Cæsarean operation succeeded a former delivery by embryulcia, as an operation of election, and all the women and children were saved. Dr. W. S. Playfair says, in his treatise on midwifery ("Science and Practice of Midwifery," American edition, 1878, page 502) : "Great as are the dangers attending craniotomy in extreme difficulty, there can be no doubt that we must perform it whenever it is practicable, and only resort to the Cæsarean section when no other means of delivery are possible." This is the generally accepted doctrine of the English school of obstetrics of the present day, although Radford, Greenhalgh, and a few others are opposed to it. Denman and Meigs questioned the propriety of repeatedly performing craniotomy in the same woman, and the latter was one of the first to act upon it in the United States, when he refused thus to deliver Mrs. Reybold in her third labour.

Dr. Playfair says : "He would be a bold man who would deliberately elect to perform the Cæsarean section on such grounds," and I am happy to answer that we have had several such bold men, and that they were repaid in a remarkable manner by success. What better trophy could Dr. Meigs, if

now living, present, than Mrs. Reybold, with her two children
and six grandchildren, as the fruits of his declining, in 1835,
to destroy any more children for her? There can be no
question now but that Mrs. Reybold not only suffered far
more in the two craniotomics, and was in more danger afterwards, than from the two operations of Professor Gibson.

The Operation.

Although I shall not attempt to describe the steps of the
operation, there are points to which it may be well to call
special attention, in view of the teachings of the past.

1. The nearer the abdominal incision is made to the central
line of the linea alba, the less will be the hæmorrhage.

2. The earlier the operation the better for the safety of the
mother and child.

3. Chloroform, by leading to uterine inertia and vomiting,
is an unsafe anæsthetic. Local anæsthesia by spraying the
line to be incised is safer.

4. The best sulphuric ether is a safer anæsthetic than
chloroform.

5. In the days before the use of anæsthetics, the Cæsarean
operation was safer than now, as there were no secondary
anæsthetic effects.

6. The operation is not very painful after the skin has been
incised; this is painful, and feels like burning with a hot wire.
The stitching is the most severe.

7. To arrest uterine hæmorrhage and prevent its return,
suture the uterus with silver wire stitches.

8. Ice is a good remedy for exciting uterine contraction,
and much safer than the persulphate or perchloride of iron.
Vinegar is also a valuable excitant, and acts promptly. Ergot
is a good preparative to avoid inertia.

9. The abdomen should be thoroughly cleared of all the
blood and amniotic fluid which have escaped from the uterus
during the operation.

10. Septic poisoning is apt to originate in the decomposition of matters that have escaped from the uterus, even when
in small quantity.

11. Many women lose their lives through post-partum
uterine relaxation, ending in hæmorrhage. To avoid this,

operate very early and without anæsthesia. In all late cases, suture the uterus with silver wire for safety.

12. Where the uterine drainage is not good, leave the lower part of the abdominal wound open, and syringe out the abdominal cavity with dilute liq. sodæ chlorinat. f℥ij to Oj, or bromo-chloralum one part to forty or fifty of warm water, daily.

13. Never use catgut for uterine sutures; as the knots become untied, the wound opens, and patient dies.

14. If the temperature of the room is high, the wound may be kept open until the uterus is safely contracted, all bleeding arrested, and parts cleansed. In one case the wound was not closed for an hour, and the patient recovered.

15. If the fœtus is dead and putrid, sponge out the uterus carefully and put five or six sutures in it. It is safer to do this than run the risk of secondary hæmorrhage, or escape of lochia into the peritoneal cavity. Two women, seven and ten days in labour, were thus saved in the United States, and are now alive and well.

CRANIOTOMY will have to be performed even when it is denounced as an operation of election, it will always have to be accepted as one of necessity, I therefore deem it right to give a short statement of the instruments which are employed in its performance.

Perforator.—This instrument ought to be well made and strong, the blades should be straight, as when curved the instrument is apt to slip. It is the first required in the operation, and it ought never to be dispensed with. It is a safe instrument, if used with care and judgment, but there is great danger if badly constructed, or when used carelessly. I once knew the projecting promontory of the sacrum pierced in mistake for the cranium.

Crotchet.—The crotchet was the only instrument used as an extractor, after perforation, for many years. It is powerful in the hands of a skilful and dexterous operator, but it is a dangerous implement both to the maternal structures and the hand of the obstetricist if used without great care. Dr. A. R. Simpson speaks very disparagingly of it, and says: " If I may borrow a comparison from the rather too popular game of war, the crotchet when compared with the cephalotribe or the cranioclast has about the same relative value for the purpose of saving the lives of poor women in labour, as the old Brown Bess has, when compared with the Schneider or Martini-Henry rifle, for killing their husbands in battle." Dr. Underhill says: " He could not quite agree with Professor Simpson in his denunciation of the crotchet, for he at least had more than once succeeded in delivering with it when he had failed with some of the other instruments, the crotchet had the great advantage of working entirely within the fœtal skull," &c. Dr. Barnes says: " Until I had contrived a good craniotomy forceps I myself trusted to it entirely."

Dr. David Davis has contrived and recommended his guarded crotchets as being much safer and more effective. There are two delineated in his "Operative Midwifery." In one, the crotchet

part is placed on the outside, and the guard on the inside of
the cranium. The other is constructed for the crotchet
portion to be placed inside and the guard on the outside of the
cranium.

Craniotomy Forceps.—Various kinds of craniotomy forceps have
been constructed. The first pair, recommended and used by Dr.
Haighton, was a pair of lithotomy forceps with teeth in the
hollow. The use that this instrument could be effectually put
to was to detach and bring away portions of bones. Holmes,
Conquest, and Barnes, have each had a preference instrument.
It is considered to be much safer, and, indeed, is better, than
the crotchet. One blade is passed within the cranium and the
other on the outside of the cranium. The use is to break up,
detach, and bring away portions of bone, and ultimately to be
so fixed and to take such hold as to be an effectual extractor.

Osteotomist.—The osteotomist is a new application of mechanical power, combining the principles of a punch and a pair
of scissors. It is very strong, and consists of two fenestrated
oval rims of unequal size, but of nearly equal strength. The
smaller is of a size to enter into and to fit closely within the
parietes of the larger. Dr. Davis also contrived another
osteotomist in which the punch-shaped foramen is oblong. He
was in the habit of showing its extraordinary power to his pupils
by making large breaches in strong ribs of beef. He says :
" If the smallest measure of pelvis, compatible with the extraction of a dead child piecemeal by the natural passages
without great risk of danger to the mother from the force used
in the operation, has been more disputed than positively determined, yet I have known examples of fœtal heads so large
(even after being opened and evacuated, as not to admit of
being brought down through the superior aperture into the
cavity of the pelvis, without the exertion of considerable force,
even when I had formed my estimate from the most accurate
admeasurement I was capable of instituting) that the conjugate diameter of the brim amounted to little less than three
inches."

If we suppose the conjugate diameter not to exceed *two
inches and a half*, the extraction of the base of the fœtal skull
will necessarily be attended with much additional difficulty.
But if the intermediate space between the symphysis of the

pubis and the promontory of the sacrum be presumed to be no more than two inches, then the attempt to extract a full-grown child through natural passages by means of the crotchet in common use, or by any crotchets used, with much force, the maternal structures would suffer from contusion, and the woman herself exposed to eventual loss of life.

"The broadest oval ring" (of the osteotomist) is precisely three-quarters of an inch. I may therefore take it for granted that wherever there may be sufficient space to admit of the introduction of the instrument, together with the point of an index finger to feed it with successive purchases of bone, it will be practicable to effect, and therefore prudent to attempt, the delivery by the natural passages. "There are few pelves even in large collections of distorted ones with superior apertures so small as not to furnish from between one inch to one inch and a half of space in the direction of their conjugate diameters, or at least of antero-posterior diameter across some part of their brim. In any such case I should think it my duty to avail myself of the use of the osteotomist, and to undertake the delivery by the natural passages. If, indeed, I am not greatly over-rating the power of this instrument, it will not only enable a skilful operator to effect deliveries in cases of moderate distortions with much more facility to themselves and proportionately less danger to their patients than heretofore, but it will also have the effect of reducing almost to zero the necessity for having recourse to that last extremity of our art, and the forlorn hope of the unhappy patient, the Cæsarean section.

"I may be here expected to give my opinion as to what should be considered the smallest space at the brim of the pelvis, compatibly with practicability of delivery by the natural passages, with the option of using the osteotomist. I feel that I cannot well answer this question without incurring more responsibility than I am willing thus categorically and without some further explanation to undertake." Dr. Davis had a series of machines, for instructing his pupils, carved in oak and made accurately to represent, as far as art can go, distorted pelves, one of these is constructed on the model of Elizabeth Thompson's pelvis. "The largest circle that can be formed in any part of the superior aperture does not exceed in diameter one inch." The inference to be drawn is that with the aid of the osteotomist the mutilated fœtus was drawn through this representative pelvis. The pelvis of Elizabeth Thompson is in the Museum of St. Mary's Hospital, Manchester, and was presented by me. There is a block of oak carved on this model which with others I have given to the aforesaid hospital. I have endeavoured to bring a mutilated infant through it, but I never succeeded. However, it is one thing to operate on an inanimate machine, a block of wood, let it be ever so accurately formed, and another to operate on the pelvis of a living woman (see Part I. page 54). I deny the possibility of bringing a mutilated full-grown child through such a pelvis, whatever appliances are used. The case of Elizabeth Thompson ought to be read and studied by every obstetrician as related by Mr. William Wood, "Medical Memoirs," vol. v., and the controversies by the late Dr. Hull, Mr. W. Simmons, Mr. Tomlinson, and others.

Cephalotribe and *Cranioclast.*—In 1829 Baudelocque Neveu proposed a cephalotribe for crushing the infantile cranium in cases of labour obstructed by distortion of the pelvis, but in Great Britain and Ireland it was not accepted, and was treated as inapplicable to fulfil the object proposed. It is a bulky and clumsy looking implement. This instrument, nevertheless, was appreciated and used in France and on the Continent. Several instruments of this kind have been contrived in these countries, one by the late Sir J. Y. Simpson, Dr. B. Hicks, Dr. M. Duncan, Dr. Kidd, and Dr. D. Lloyd Roberts. The cephalotribe is a large, strong, and powerful instrument, and requires for its safe use a sufficiency of pelvic space, and doubtless would be much easier and more safely applied in a pelvis distorted by rickets than one distorted by mollities ossium. Obstetricians differ as to the amount of pelvic space required for its safe introduction. The minimum measurement of the conjugate diameter through which a mutilated infant could be brought, as stated by some obstetricians, I should say, would be most dangerous.

Dr. A. R. Simpson, Professor of Medicine, Edinburgh University, says : " I confess I have not been able to come to a definite decision as to the relative value of the cephalotribe and the cranioclast. The inevitable elongation of the head in

a direction opposed to that in which it is compressed by the
blades always seems to me a serious drawback to the value of
the cephalotribe. The extraction of the head, in accordance
with a scientific idea of the mechanism, seems in such a condition next to impossible. Hence I have been inclined always
first to try the cranioclast, but have found that when the head
was very movable above the brim the cephalotribe took hold
and crushed it where it was more difficult to get a satisfactory
grasp with the cranioclast. At the same time, I believe the
obstetricians are right who hold that the cranioclast is applicable, and may be successfully employed in extracting the
head through a pelvis too contracted to admit of the use of
the cephalotribe."—*Edinburgh Medical Journal*, April, 1880,
page 869.

" When the brim was thus contracted, the problem was how
to reduce the head so as to minimize the danger to the mother.
With the cranioclast and cephalotribe they could undoubtedly
pull the head through the pelvis, but very often this had to be
done with a force exhausting to the practitioner's muscle, and
damaging to the mother's tissues. The head, more particularly when it was grasped with the cephalotribe, was always
elongated in one diameter, and laceration of the maternal
structures inevitably ensued. In this way the cephalotribe
was specially dangerous.

" It was really striking that an obstetrician of Karl Braun's
(of Vienna) experience should pronounce so strongly in favour
of the cranioclast. Braun's opinion, and the Paper on his
results with it by Karl Rokitansky, had led to its general
adoption in Germany. Dr. Gordon seemed to think the
cephalotribe a finality. They were a long, long way from that,
as the many forms employed clearly showed. He remembered
the unfavourable impression as to Dr. Duncan's cephalotribe
which he received on reading Dr. Duncan's Paper, and which
was confirmed by the condemnatory criticism of Dr. Mac-
donald," &c.—*Edinburgh Medical Journal*, May, 1880,
page 1,033.

Dr. Keiller disapproved of the objections expressed by Professor Simpson against the power and value of the cephalotribe.
" While expressing his faith in the power of a well constructed
and properly handled cephalotribe, which he had on yman

trying occasions tested, and as his exhibited series of cephalotripsy preparations—the Hibernian 'army of martyrs'—fully
proved; it was, he held, a mistake to imagine, as the Professor
seemed to do, that it was necessary to attack so as to completely crush the bones of the face of the fœtal skull. He had, as
yet, no reason to change his entertained views and teachings in
regard to this most important practical matter, and confessed
that the ponderosity and power displayed in some of our
cephalotribes and other midwifery engines were such as to
have often led him to remark that they might serve the very
unequal purpose of 'quarrying stones and crushing babies'
bones.' He had been much disappointed in the operation of
cranioclasm, and mentioned his special objections to the
cranioclast as a craniotomy instrument."

We learn from the opinions expressed by these two eminent
obstetricists, that the capabilities of the cephalotribe and the
cranioclast are not settled as implements to bring a mutilated
child through a distorted pelvis. I confess I find, in the discussion on the obstetric value of either one or the other
instrument, no mention made of the minimum space in the
antero-posterior diameter of a distorted pelvis, through which
a mutilated child can be brought by the aid of each of these
instruments.

Vertebral Hook.—Dr. Oldham's hook is a slender hook, with
the hook portion standing off rather acutely; it is appropriately
applied in extracting the head when separated from the trunk
by being passed into the vertebral canal.

Wire Ecraseur.—Dr. Barnes has proposed the application
of a wire ecraseur, for the purpose of reducing the bulk of
the cranium of the child in cases of extremely distorted pelves.
He says: " I also demonstrated this operation at the meeting
of the Obstetrical Society, of the 2nd of June, 1869. This
demonstration was on a delicate, rickety pelvis, measuring an
inch in the antero-posterior diameter, and scarcely more in the
sacro-cotyloïd diameter." This experiment was made on a
dried rickety pelvis, in which there were no tissues liable to
meet with injury, and in which operation two senses, the eye
and the hand, could be advantageously used for accomplishing
the object intended. The result of such an operation on a
woman in labour, having an extremely distorted pelvis from

mollities ossium, would be very different.   The head has to be
perforated, and the crotchet used to steady the head, so that the
pelvic structures must necessarily be subject to considerable contusion, and, I should say, the operation, to complete failure.

Professor Simpson says: "Dr. Barnes has proposed to
reduce the head by passing round it the loop of a wire ecraseur,
and so cutting it through from above. But my impression
regarding this suggestion has been, that in cases where it
would be possible to apply a wire round the head of the child
it would be easier to reduce it with the cephalotribe or the
cranioclast.

Basilyst.—Professor Simpson has proposed an instrument,
which he has named Basilyst, for breaking up the fœtal head.
He first brought out his proposition in a Paper which was read
and discussed at the meeting of the Obstetrical Society of
Edinburgh, held January 28, 1880.   The discussion is reported
in the *Edinburgh Medical Journal*, May, 1880, page 1,030.   The
Paper is published *in extenso* in the *Edinburgh Medical Journal*,
April, 1880, page 865.   He appears to have derived the impression of the necessity of drilling holes in the basis cranii
from G. M. Guyon, " which he described as intercranial
cephalotripsy, and had successfully put in practice at the Hôtel
Dieu."   He possessed a case containing Guyon's apparatus,
lately belonging to the late Sir J. Y. Simpson.   He used this
apparatus in the delivery of a slender-built woman, with a
universally and unequally contracted pelvis.   Baudelocque's
diameter being $5\frac{1}{2}$ inches ; conjugate diameter at the brim of $2\frac{1}{4}$
or $2\frac{1}{2}$ inches. When two holes had been drilled in the base with
Guyon's perforator, and the head compressed between the
blades, he was astonished how easily the delivery was effected.

Dr. Keiller said : " There was seldom much good resulting
from multiplicity, and far less from complexity, in operative
measures ; for, as a rule, the more simple and handy our artificial aid in cases of emergency is made, the more useful and
the more procurable it will in all probability be found.   This
operation of so-called ' Basilysis,' this spearing and com-
minuting of parts, which seldom or never seriously interferes
with the termination of labour, seemed uncalled for."   It was
considered that Guyon's instrument was cumbersome.

Professor Simpson showed his Basilyst at a meeting of the

Obstetrical Society, February 25, 1880. The *Edinburgh Medical Journal*, page 1,189, says : " So satisfied with it was he, that he ventured to demonstrate its use on a dead fœtus before a class of 170 students. The pelvis he employed was a rickety one, with a conjugate of less than $2\frac{1}{2}$ inches, so that there was some difficulty in extracting the shoulder."

An experiment of this sort would be very easily performed, with no structures to suffer, and with the whole matter open before him, and the cavity of the pelvis roomy, and the conjugate diameter of the brim so much wider than what exists in many rickety pelves ; doubtless the dangers and difficulties would be much greater in pelves distorted by malacosteon.

CASES.

I HERE offer to the reader the following seven cases of Cæsarean section which had been previously published. I have preferred giving them as they originally appeared, rather than making any abbreviations of them, although many remarks which are appended to them will be found to have been already made in the preceding observations, &c. I have affixed to each case the number under which it is registered in the Table. One case is now published for the first time.

I have given outline sketches of the brims of most of the pelves belonging to the women whose cases are hereafter related. Those measurements which were obtained after death (with the exception of that belonging to Mary Haigh) were made after all the soft parts had been removed, so that they (the measurements) are really greater than they would have been found, with the addition of the pelvic organs and tissues.

I have given two delineations of each of the pelves of the cases Nos. (Tab.) 49 and 53. One shows as accurately as possible the state of the pelvis when the operation was performed; the other indicates the degree of diminution which had taken place after this time up to the death of each of the patients. The cavities and outlets of all the pelves were relatively quite as much diminished as the brims are represented to have been.

CASE (Tab.) 27 (*Unsuccessful*).—On Sunday the 1st of April, 1820, I was requested to visit Mary Ashworth, residing at Denton, about six miles from Manchester. I was told she was in great danger, having been in labour a considerable length of time, and that no progress was made in the case. This report did not surprise me when I ascertained who the individual was; for Mr. Wood, my partner and esteemed relative, had visited her about the end of the seventh month of her present pregnancy, at the request of her medical attendant, Mr. Morris, a highly respectable surgeon, who resided at Ashton-under-Lyne. Mr. Wood at this period examined her *per vaginam*, and

his opinion was that if her pregnancy did proceed, when labour
came on the Cæsarean section would be required, as in her
case no other means would be of the least avail. At three
o'clock, P.M., I reached her dwelling, and found Mr. Morris
and Mr. Cheetham awaiting my arrival.

I was informed by Mr. Morris that she had been in strong
labour about thirty-four hours; that the membranes had
ruptured in two hours after its commencement, and that the
liquor amnii had gradually passed away. He had not been
able to feel the presentation nor the os uteri. The pains were
strong for twenty-four hours, but afterwards gradually abated.
The urine had been passed freely during the Saturday, but this
day (Sunday) there was no evidence of any having been discharged. The bowels were constipated, and had not been
opened during the labour.

Her previous history was to the following effect :—She had
borne ten children, nine of whom were expelled by the natural
powers. In the last labour considerable difficulty occurred,
and the practitioner had recourse to craniotomy. During her
tenth pregnancy she experienced considerable weakness in her
loins, and felt rheumatic pains about the hips, and limped in
her gait. These pains continued from that time till her present
pregnancy, but did not increase in degree. When she again
became pregnant her sufferings increased, and her lameness
became more manifest. Her stature was now observed to
diminish in height. She was forty-two years of age, and was
employed as a hat-trimmer.

I found her in bed, lying upon the back, with the head and
shoulders raised. She moved with the greatest difficulty. The
pulse was feeble and frequent, beating about 150 in the minute.
She had often vomited, and had great tenderness in the belly,
which was considerably increased by pressure. Her tongue
was furred and dry, and she complained of great thirst; her
countenance expressed considerable anguish. Being requested
to compose her mind, she answered : " She was composed, but
anxious for relief, and would suffer any pain so that she might
be delivered."

Upon examining the abdomen I found the uterus projecting
very much forwards, and lying with its anterior surface upon
the upper part of the thighs. By a vaginal examination I

K

discovered that the labia were much swelled, and the vagina felt dry and rough; it was hotter than natural, and an odour similar to that arising from animal matter when partially decomposed was perceived from the hand when it was withdrawn. The outlet of the pelvis had undergone great change; the arch of the pubes was totally destroyed by the near approximation of the rami of the ischia and pubes, having only a small slit, so narrow at the upper and lower parts as not to admit the point of the index finger; at the middle, however, the finger could just be introduced. The tuberosities of the ischia were not more than one inch and a half to one inch and three-quarters as under; and the lower portion of the sacrum was so much more incurvated than natural as to throw the coccyx much more forward, and consequently lessen the conjugate diameter of the lower aperture of the pelvis. This great diminution in the outlet rendered it difficult to pass the hand in order to measure the brim, and it was found necessary to carry it very far backward in order to accomplish it. This aperture was found much more altered than the outlet; one finger only edgeways could be placed between the points of bone in the conjugate diameter. In traversing it from side to side I could detect no great difference, but if there was any, the left was the most contracted. In the transverse diameter I could just place three fingers parallel to each other. The figure* of the brim was tripartite, having a slit on each side and a third passing forwards, produced by the approximation and jutting out of the pubes, which was so narrow that the finger could not pass within it. This alteration in the brim was occasioned by the falling downwards and forwards of the upper part of the sacrum, and the lower lumbar vertebra, and by the body of the ossa pubis and ischii being forced backwards and inwards, whilst the symphyses and rami of the pubes projected forwards and upwards. The measurement of the conjugate diameter did not exceed three-quarters of an inch, or that of the transverse two inches; and having placed my fingers upon each other in the widest part, and having measured them when withdrawn, I concluded that no body of a diameter greater than from three-quarters to an inch could pass through it, and that delivery *per vias naturales,* aided by the crotchet,

* See Fig. 1st.

was utterly impracticable. Another important feature in the case was, that no part of the child or os uteri could be felt.

Upon these grounds, then, we concluded that our only resource was the Cæsarean section. Our opinions were now stated to the husband and friends, and they cheerfully submitted to any practice we thought best to adopt. The patient, anxious to have her sufferings terminated, also readily acquiesced in our decision.

An enema was ordered to be administered, and it soon operated. The catheter was also introduced, but little urine was withdrawn. As the patient felt cold, a little warm wine and water was given, which acted beneficially. Having placed her upon a table, an incision six inches long was made through the abdominal integuments, about one inch to the left of the umbilicus, extending from three inches above to three below. A small opening was made into the peritoneum, and this membrane was afterwards fully divided by a probe-pointed bistoury. The uterus was now exposed, and an excision of equal length was made into this organ, nearly dividing its entire substance. An opening was now made at the lowest point of the wound by the knife, so as to admit the finger, upon which the bistoury was again passed, and the uterus was laid open. I now passed my hand, and, taking hold of the thigh of the child, readily extracted it; but, unfortunately, it was dead. The funis having been divided, the placental portion was held firmly in the left hand, whilst the right was introduced into the uterus to extract the placenta, which was attached to the upper and posterior part of the uterus. As soon as the placenta was removed the uterus energetically contracted, and lowering itself became almost invisible. The intestines protruded at the wound, but were soon reduced and retained by the hands extended over their surface. Mr. Morris next passed several ligatures through the abdominal parietes, and afterwards applied slips of adhesive plaster, by which the edges of the wound were closely approximated. Pledgets of lint, spread with cerate, were also applied, and in order to secure the whole, a broad bandage was loosely put on. The quantity of blood lost was trivial, not exceeding three or four ounces, which favourable circumstance, doubtless, was partly owing to the position of the placenta, and partly to the vigorous contraction of the uterus.

During the whole course of the operation the patient maintained the greatest fortitude, and expressed her thankfulness upon the termination of her sufferings. She, however, as well as all present, was disappointed that the child was lost.

The patient was then put to bed, and as the pulse was rather low a cordial was administered, which in a little while revived her. An anodyne draught, containing sixty minims of laudanum, was also given.

At ten o'clock P.M. the pulse was 140 ; the skin was hot ; she was thirsty, and complained of headache ; belly tender ; discharge not more than usual ; thinks she can sleep ; no urine passed.

April 2nd, eight o'clock A.M.—Says she has slept ; has taken some refreshment ; skin hotter ; pulse 140 to 150 in the minute, and feels sharp to the finger ; has not voided any urine ; bowels not moved. Ordered saline effervescing draughts. The catheter to be introduced, by which from three to four ounces of urine were withdrawn.

Twelve o'clock noon.—Not so well ; had shivering and some vomiting ; the pulse was more frequent, and the abdominal tenderness increased ; belly swelled ; a sanious discharge was oozing from the wound ; bandage uncomfortable ; vaginal discharge rather greater and more offensive ; bowels not moved. The medicines were continued, and the bandage loosened.

Four o'clock P.M.—Has again shivered ; continues to vomit ; pulse more frequent and tremulous ; countenance more depressed ; abdomen more tender and more swelled ; is very thirsty, and her tongue is very much loaded ; the bowels still constipated. The saline medicines were continued, and an enema, with oil of turpentine and castor oil, was ordered to be administered.

Ten o'clock P.M.—Vomiting continues unabated ; pulse still very frequent and much weaker ; skin colder and rather clammy ; is slightly incoherent ; belly very tender and much swelled ; discharge offensive ; has not passed urine ; the enema operated. The catheter introduced, and four ounces of urine withdrawn ; the bandage still further loosened ; to have a little brandy in her gruel; to take forty minims of laudanum. The symptoms continued to grow more unfavourable during the night, and she died at four o'clock this morning (Tuesday) about thirty-five hours after the operation.

An application was made to examine the body, but permission was only granted under a promise that the wound was alone to be inspected; but Mr. Morris, while alone, took the opportunity of ascertaining, as far as he was able, the state of the parts. The edges of the external wound were quite separate, and had a flabby, unhealthy aspect. Having divided the stitches and drawn aside the integuments, the uterus was observed to be well contracted. The wound was much diminished, its edges were loose and unhealthy. The peritoneum was inflamed, and about from four to six ounces of serum were effused within its cavity. Upon again raising the uterus, the cervix was seen to be dark-coloured, and having divided it, the lower portion and orifice were found in a gangrenous state. The bladder was empty, and uninjured. The brim of the pelvis was examined, and found fully as much distorted as I have before mentioned.

Remarks.—The issue of this case prevents indisputable evidence of the serious mischief arising from protracting the operation. Both the life of the mother and the child were most likely forfeited by the delay. How the real character of the case could have been so much overlooked, after the clear and decided opinion of Mr. Wood, given at the end of the seventh month, I am at a loss to conceive.

The tumefaction of the external genitals, and the inflamed condition of the vagina, are alone to be attributed to the too frequent examinations made. When a practitioner undertakes to explore for the exact measurements of the brim of the pelvis, deformed like the one belonging to the subject of this case, he is compelled to pass his hand completely into the vagina, and, from an anxiety to accomplish this, and to ascertain the nature of the presentation of the child, he is induced to repeat the operation very often. These repeated examinations are often productive of very serious mischief, causing inflammation, which frequently terminates in suppuration and sloughing. With these circumstances before us, we are of opinion that every unnecessary manœuvre ought to be avoided, and that the practitioner should acquaint himself, as completely as it is possible, with the nature of the case before he withdraws his hand.

CASE (Tab.) 30 (*Unsuccessful*).—Mary Nixon, aged thirty-nine, had been married sixteen years, and had been pregnant eight times, in seven of which she reached the full period of gestation, and in one she miscarried once at the fourth month, which happened about thirteen months before her present pregnancy. The last natural labour took place about four years since, and was so rapid as to be completed in two or three hours. She had enjoyed good health until about two years ago, when she began to suffer from what she called rheumatism and a short cough. Afterwards she was frequently confined to bed, and her friends observed her to rather diminish in height. The pains in her back and hips increased in violence during her present pregnancy, and her height is now very considerably diminished. She has been employed as an "ender and mender" for the manufacturers, which occupation has obliged her to be sedentary, but has attended to her domestic duties, although unfit.

At one o'clock on Thursday morning, May 24, 1821, she was apprised of the approach of labour by a discharge of water, which continued to dribble away without pain. At four o'clock, Mrs. Barber, her midwife, was sent for, as she now felt slight pains. On examination *per vaginam*, Mrs. B. could neither feel the os uteri nor any part of the child, but ascertained that the pelvis was considerably distorted. At noon she sent for Mr. Wilson, one of the surgeons of the Manchester Lying-in Hospital, but he was from home, and instead of immediately applying elsewhere, she allowed several hours to elapse before she sent for other assistance. At eight o'clock P.M. Mr. K. Wood saw the patient, and considered the case of such importance as to induce him to call upon Mr. Wilson, who was then at home, and they immediately went to the house, and reached it at a quarter before nine o'clock. Mr. Wilson agreed in his opinion, and he desired that a general consultation of the medical officers of the institution might be immediately called.

At ten o'clock P.M., when Dr. Hull, Mr. Wilson, Mr. K.

Wood, Mr. Lowe, and Dr. Radford had assembled, the state
of the patient was as follows:—Her pulse was 130 ; the skin
hotter than natural; her tongue was furred; she was very
thirsty ; her countenance was cheerful ; she had passed urine
at several intervals during her labour, and her bowels had
responded three times to an enema which her midwife had
very judiciously administered. The stools were scanty, and
of a green colour. The pains, which were reported to have
been very frequent during the afternoon, continued so. She
complained of great tenderness in the belly, which was considerably increased by pressure. The distance between the
pubes and sternum was much shorter than natural. The
fundus uteri projected very much forwards, and had an
inclination to the left side. By an examination *per vaginam*,
I found the parts soft, moist, and cool. The sacrum was considerably more incurvated than natural, and the coccyx projected upwards into the cavity of the pelvis. The tuberosities
of the ischia approached very near together at the fore-part,
and the rami of the ischia and pubes approximated so closely
together as not to admit a finger to pass between them along
any part, except at the middle, at which place there was a
small opening, in consequence of a slight bulging outwards of
the bone on either side. The pubic arch was destroyed, and
only a small chink left, by which the depth of the pelvis was
increased at the anterior part. In order to examine the brim,
I passed my hand, but was compelled to carry it very much
backwards. The pubes on each side formed a very acute
angle at their fore part, and then running forwards nearly
parallel to the symphysis, having a slit between them which
would barely admit the finger edgeways. The base of the
sacrum and the last lumbar vertebræ had sunk forwards and
downwards into the pelvis, and diminished the conjugate
diameter on the left side so much as barely to admit the finger
in the position, that when it was withdrawn, it measured
three-quarters of an inch, and when placed in other parts of
the brim, an inch, as far as could be ascertained, was the
fullest latitude which could be given to guide us in our
decision.* We all agreed that no other means but the
Cæsarean section could avail us to deliver this poor creature.

* See Fig. 2nd.

Having decided upon our plan, we stated our opinions to the friends, who readily consented that we should adopt any practice we thought best. When we acquainted the patient with the difficulty of her case, and the operation necessary to extricate the child, she unhesitatingly acquiesced. It was intended to have used the catheter, but this was unnecessary, as half a pint of urine was discharged by her own efforts.

She was now placed upon a table, and a little brandy and water, with thirty drops of laudanum, was administered her.

Mr. Wilson made a longitudinal incision a little to the left of the umbilicus, six inches in length, extending from three inches above to three inches below this part, and divided the abdominal parietes down to the peritoneum. A small opening was made through this membrane, and it was fully divided by a probe-pointed bistoury, passed along with the finger. An incision was now made nearly through the uterus, corresponding in length and direction to the external wound. The probe-pointed bistoury was introduced on the finger through a small opening, and the remaining portion divided. This exposed the child, which lay with its breech towards the opening. Mr. K. Wood seized the child by one thigh, and the body was extracted with the greatest ease, until the shoulders came to pass, when the uterus suddenly and powerfully contracted, and grasped the child's neck and left arm so strongly that this gentleman could not liberate it, although he used great force in extraction. He then gradually passed his hand along the body of the child into the uterus, and having dilated the structure the child was extracted. It would have been easier to have torn away the uterus from its connections than to have brought the child away by direct extractive force. The fundus and body of the uterus felt very hard. The child was vigorously alive when first taken hold of, but from the length of time occupied in extracting the head, it became so enfeebled as to show only slight signs of life. I very diligently employed every means to resuscitate it, and continued them for at least three-quarters of an hour, but was ultimately unsuccessful. This was a most appalling affair. After dividing the funis, the placental extremity was firmly held with one hand, whilst the other was introduced into the cavity of the uterus, for the purpose of removing the placenta, which was already detached and lying loose. The uterus then immediately fully contracted.

The intestines, which appeared at the wound, were replaced and retained by the extended hand; the edges of the wound were then brought together by ligatures, supported by strips of adhesive plaster and an extended bandage.

Very little blood was lost during the operation, a small branch of the epigastric only being divided. Its bleeding was restrained by the pressure of the finger.

The patient felt faint whilst on the table, but was soon recruited by taking a little brandy and water. When all was adjusted she was carried to bed, and said she was quite as comfortable as she could possibly expect. The pulse now beat about 136 in the minute, and was distinct. The heat of the skin was not much above natural. In half an hour afterwards she felt a distressing sensation at the chest; her heart beat very quickly, and the breathing became very much hurried; her skin grew cold, and the vaginal discharge was increased, but still not in such a quantity as to create alarm. Thirty drops of laudanum in a little brandy and water were immediately administered, and in half an hour forty drops more. In a very short time all these symptoms subsided, and she felt as well and as warm as before. All stimulants were now prohibited, and the antiphlogistic regimen recommended, and she was left for the night in the charge of Mr. Hunt, at that time a pupil of Mr. Wilson.

Friday, May 25, seven o'clock, A.M.—Present, Dr. Hull, Mr. Wilson, Mr. K. Wood, and Dr. Radford. She experienced no further palpitation of the heart; slept tolerably well; the pulse was 131; respiration easy; skin rather hot; belly feels comfortable and not swelled. To take a saline effervescent draught every three hours, and an ounce of the almond mixture with five drops of laudanum in the intervals.

Twelve o'clock, noon.—Present, Mr. Wilson, Mr. Hudson, Mr. K. Wood, and Dr. Radford. The heart appeared to jerk; pulse 130, and quite distinct; skin hotter; her countenance more anxious; tongue furred, but moist; has again slept; urine passed twice. The medicines were continued.

Four o'clock, P.M.—Present, Mr. Wood, Mr. Wilson, Mr. K. Wood, and Dr. Radford. Belly rather tense; pulse 130, and firmer; tongue dry and furred; is thirsty; has passed urine; her cough is still troublesome; the bowels are constipated. A solution of Epsom salts in infusion of roses was directed to be given until it operated, and a linctus for the cough was ordered.

Six o'clock P.M.—Present, Dr. Hull, Mr. Wilson, and Dr. Radford. Her countenance looked better; the heart throbbed violently; her pulse beat 125; the tongue was rather more moist and soft; the belly continued very tense, and the respiration was hurried; she has had slight vomiting, and her bowels have not yet been moved. The medicines were continued.

Ten o'clock P.M.—Present, Mr. Wilson, Mr. Lowe, and Dr. Radford. Her countenance has become more anxious; her respiration is more laborious, and she has again vomited. The skin is hotter; her belly is very tender, and is much swelled; her pulse is 130; the vaginal discharge trifling, and very slightly coloured; she has a tendency to doze; has had several fœtid liquid stools.

To omit the aperient, but to continue the other medicine; the bandage to be loosened.

26th, Saturday, eight o'clock, A.M.—Present, Mr. Wood, Mr. Wilson, Mr. K. Wood, and Dr. Radford. Has frequently vomited a brown slimy fluid; her pulse is 130; her respiration still laborious; the belly rather softer; her skin is still hot; and the lochial discharge very trifling.

Ten o'clock A.M.—Present, Dr. Hull, Mr. Wilson, and Dr. Radford. Pulse 134; vomiting has ceased.

One o'clock P.M.—Present, Mr. Wood, Mr. Wilson, and Dr. Radford. Her hands feel cold; the pulse is 130; her mind is clear; the vomiting has ceased; lochial discharge fœtid and more profuse; and there has been a thin and offensive sanious discharge from the wound.

Five o'clock P.M.—Symptoms still grow worse.

Ten o'clock P.M.—Present, Mr. Wood, Mr. K. Wood, and Dr. Radford. The symptoms continue to become more unfavourable.

Sunday, eight o'clock A.M.—Pulse 140, and very weak; the skin is rather cold, and covered with a slight clammy sweat; her countenance is very anxious, and there is great swelling and tenderness of the belly; the lochial discharge is very offensive; she has not passed urine. The lowest strap of adhesive plaster being removed, the wound appeared in an unhealthy state and not united; a great discharge took place, which was very offensive. During this day (Sunday) she was visited several times, and found still further sinking. At six o'clock in the evening she expired, having lived sixty-seven hours and a half after the operation.

Remarks.—It may appear strange that no notice was taken of this poor woman's case at an earlier period of pregnancy, as she was a patient of the Lying-in Hospital. But our hospital extends its aid only to poor women at their own houses; and this poor creature having obtained a note of recommendation from a subscriber, was admitted, the medical officers having no knowledge that such a case was on the books. Another unfortunate circumstance was the midwife's omitting to send for other surgical assistance in the absence of Mr. Wilson, thereby allowing several valuable hours to elapse.

The violent contraction of the uterus, by which the head and left arm of the child were seized after the extraction of the trunk and lower extremities, forms a remarkable feature in the case. The placenta was found detached and lying loose in the cavity of the uterus; and how far this violent contraction depended upon this circumstance is difficult to say. In natural labour we well know that, as soon as the placenta is detached, the energies of the fundus and body of the uterus are aroused, and contraction follows, and is continued until this mass is expelled. In the former case of Cæsarean operation the uterus was quiescent, until the placenta was detached by the hand, when contraction instantaneously followed.

Case (Tab.) 41* (*Unsuccessful*).—On Tuesday evening, February 21st, 1841, Mr. Dunn, surgeon, of Manchester, and Mr. Goulden, of Stockport, called upon me to request my

* This (No. 41) case is reprinted from the *Provincial Medical and Surgical Journal*, vol. xv. p. 18. It was also originally published in the *London Medical Gazette*, vol. xlvii. p. 801, 1851, in which a great mistake has been made by the printers, in misplacing some of the first sheets of the manuscript, and which has created great confusion in the meaning. I have adverted to this error in order that it may be, when possible, in some measure corrected.

opinion on the case of Mary Forrest, who resided at Stockport. She was the patient of the latter-named gentleman, from whom I learnt the following particulars:—He had visited her at one o'clock P.M. (on Monday) the day before. She was at the full period of gestation, and had violent and regular pains in the back. Mr. Goulden made a vaginal examination, and found the pelvis very much distorted; he thought she could not be delivered unless by the Cæsarean section. A grain of opium was ordered to be taken every two or three hours until the pains abated, which happened after six doses had been taken. When Mr. G. visited her on Tuesday morning he found her nearly free from pain. Her bowels being constipated, some pills composed of aloes and calomel were prescribed. He again examined her *per vaginam*, and concluded that her delivery was impossible, except by the means already stated. He further said on the evening of this day (Tuesday), when he was at my house, that she was now entirely free from pain, and, on inquiry, said he did not think labour had commenced. The object of his coming over to Manchester was to request me to visit the patient the following day and examine the pelvis; so that clear and definite views might be had, and what course should be adopted when labour came on. On Wednesday, I accompanied Mr. Dunn to Stockport. We found Mary Forrest in bed, and so helpless as to be unable to turn herself. Her countenance was short and inexpressive; her lips thick; her eyes blue; on the corner of the right eye there was a considerable opacity, the result of an injury from having been struck by a shuttle. She had been employed as a power-loom weaver in a cotton factory. Her age was thirty-eight years. Her first labour, which was natural, happened nineteen years since, and the child was a boy. About eighteen months before her present labour she became pregnant a second time, and aborted at the end of two months. She had suffered during the last five or six years violent rheumatic pains, but otherwise her health had been good until within the last few months. Her stature had become considerably less, and her legs and thighs were very œdematous. She said she felt the movements of the infant, which, on first placing my hand on the abdomen, I thought I perceived; but, keeping it steadily applied, I soon found it was only slight contraction of the uterus. The stethoscope was used, but gave no evidence of its being alive. The abdominal parietes were very thin and felt very flabby. The uterus projected very much forward, and lay over the symphysis pubis to such a degree that when the patient was placed in a sitting posture in the bed its anterior surface rested on her thighs. Her breech was gibbous, and her thighs approximated considerably more than natural. The discharge from the vagina was of a dark colour, and had a putrid smell.

Her constitutional condition was very bad. Pulse 140 ; tongue furred ; great thirst ; continued vomiting ; great abdominal tenderness and tension ; bowels not moved for two days. By a vaginal examination the pelvis was found to be considerably distorted. The outlet was so much diminished as to render it difficult to pass the hand, in order to ascertain as precisely as possible the measurement of its several parts. The arch of the pubes was nearly destroyed by the near approximation of the rami of the ischia and pubes, there being only a narrow slit, which just admitted the point of the finger. Between the tuberosities of the ischia I could barely place three fingers. The coccyx was considerably incurvated upwards and forwards. The brim was tripartite in figure, or composed of three divisions. This altered shape was produced by the upper portion of the sacrum, and the last lumbar vertebræ standing so much forwards and downwards, inclining a little more to the right side ; and by the body of the ossa pubis and ischii being forced backwards and inwards, and the rami and symphysis pubis jutting forwards. On the right side I could just place the point of the index finger, and on the left I could pass two fingers, one lying partially over the other.* Neither the presentation of the child nor the os uteri could be felt.

Although my attendance was specially requested to explore the pelvis in anticipation of labour, and conclude on the measures which must be adopted when it supervened, we were compelled now to decide on her delivery, as she was and had been in labour since Monday, when the membranes must have ruptured, as I ascertained she had a dribbling watery discharge afterwards. We unanimously agreed that she could only be delivered by the Cæsarean section. When we apprised this

* See Fig. 3rd.

poor creature of our decision, she readily consented to submit to any plan we considered necessary. An enema was administered, but it did not act. The catheter was passed and about an ounce of urine withdrawn. Afterwards I placed the hand on the abdomen, and found still a fluctuant tumour, which induced me to pass a longer catheter, and, by a careful manipulation, it was carried higher, and succeeded in further removing a considerable quantity of urine.

In the first place, we arranged the duty which each gentleman had to perform. Mr. Cheetham and Mr. Pigott joined us to witness the operation. I commenced to make the incision a little above the umbilicus, on the left side of the linea alba, and extended it downwards towards the pubes, so as to be about six inches in length. In consequence of the very attenuated state of the abdominal parietes, the uterus was at once exposed, and its tissues partially cut. It was necessary to raise the defected uterus upwards before the external incision was made. About a pint of water, with a large portion of intestines, now escaped. This portion of bowel, which had been dangerously exposed to be wounded, being drawn aside by Mr. Dunn, I fully divided the uterus in the line of the wound first made into it. The placenta was now exposed, the greater part of which lay to the right side of the incision and maintained its adherence; whilst that on the left (the smaller division) was separated by the contraction of the uterus. Mr. Dunn immediately passed his right hand in search of the feet of the infant, and his left along the body to guard its neck from being spasmodically seized during its abstraction (a caution given to him by me, having, in a previous case, met with this disaster). Having gained a firm hold of the feet of the infant, he very cautiously and expeditiously drew the body forth until the neck came to pass, which was then firmly grasped by the uterus, and the head thereby detained. The womb had an appearance as if it was indented, and strongly reminded us of its condition in hour-glass contraction. The difficulty at last being overcome, the infant (dead and partially putrid) was completely removed. The adherent portion of the placenta was detached, and wholly extracted. Not more than a few ounces of blood were lost. The intestines were much distended with air, and were difficult to keep in position during the operation;

and could not be replaced or retained after it, until the wound was contracted with sutures. Their peritoneal coat was highly inflamed. The edges of the wound were brought together by interrupted sutures, and maintained by strips of adhesive plaister, and supported by a bandage placed round the patient. She bore the operation with great fortitude, and, indeed, scarcely made a complaint during its performance. Afterwards, she said she had not suffered so much pain as she had felt from some of the unavailing parturient pains. She now expressed herself as being very comfortable.

Nine o'clock P.M.—Considerably worse : pulse 140 ; tongue furred ; great thirst ; vomits continually ; surface cool ; countenance anxious ; great tenderness of the belly ; clyster not returned ; had no sleep ; a little urine passed. Ordered to take morph. acet. gr. j.

Thursday, 23rd, two o'clock A.M.—Mr. Goulden visited her. Pulse 160 ; vomiting unabated ; matter rejected of a dark colour ; skin rather warm ; breathing quick ; belly very tender ; had no sleep ; bowels still constipated. Ordered morph. acet. gr. j.

Nine o'clock A.M.—Mr. Goulden and Mr. Cheetham reported that she continued to vomit the same fluid ; skin warmer ; pulse less frequent, and countenance less anxious ; bowels moved ; passed urine ; lochial discharge ; abdomen tense and sore. To take morph. acet. gr. j, and saline effervescing mixture.

Half-past one o'clock P.M.—Mr. Goulden was urgently summoned : it was stated that she was dying. He found her very drowsy or lethargic ; had great difficulty to rouse her. Pulse 140 ; countenance more flushed, and rather bluish in colour ; no vomiting ; abdomen distended.

Two o'clock P.M.—Present, Dr. Radford, Mr. Goulden, Mr. Dunn, and Mr. Pigott. She still continues very drowsy ; pulse 140, and very tremulous ; skin bedewed with cold perspiration ; her countenance is sunken, and her cheeks are of a purplish colour ; her breathing very hurried, with mucous râle ; does not vomit ; belly tympanitic ; bowels not moved. Ordered an enema with spt. terebinth., &c.

The symptoms continued to grow worse until seven o'clock (vespere), when she expired, having lived about twenty-seven hours after the operation. Body examined twenty hours after death by Mr. Dunn and Mr. Gaskell, in the presence of Dr. Radford, Mr. Goulden, Mr. Pigott, and Mr. H. Winterbottom. General tumefaction of the abdomen, its parietes being extremely thin. The edges of the external wound were nearly altogether adherent; adhesion of the peritoneal surfaces of the intestines and that of the abdominal parietes. On breaking through these adhesions about a pint of bloody serum was discharged. The convolutions of the bowels were firmly united, and effused lymph was seen on their surfaces. The colon adhered to the uterus.

We were particularly struck by the very remarkable pallidity of the tissue of the intestines, when compared with their highly injected vascular state seen during the operation. The uterus stood obliquely; the wound in this organ was three and a half inches long; its edges were flabby and gaping; the os was very patulous, and its lips ragged and gangrenous. The lower portion of the cervix was very soft and dark coloured. The internal surface of the uterus was softer and darker in colour than usual. The gall bladder large and full; spleen healthy. The left kidney externally natural; its pelvis larger; and it contained a creamy fluid. Right kidney smaller than usual, and contained a similar fluid. The legs were œdematous. Pelvis removed; brim tripartite in shape; found an excessive quantity of fat about the pelvis and back.

From upper edge of the fifth lumbar vertebra to the inner surface of the symphysis pubis two inches and six-eighths.

From side of the said vertebra to the ilium behind the acetabulum on the left side, five-eighths of an inch.

Ditto on right side, one inch and one-eighth.

From the inner surface of the pubes, anteriorly, to the acetabulum on the right side, three inches and seven-eighths.

Ditto, left side, three inches and four-eighths.

Cross measure of the widest portion of the anterior slit, five-eighths of an inch.

Ditto at the point of the widest portion of the anterier slit, two-eighths of an inch.

Outlet—From the point of the coccyx to the lower edge of the symphysis pubis, two inches and five-eighths.

From the widest part of the anterior slit to the coccyx, one inch and two-eighths.

Depth of the anterior slit, two inches.

Between the anterior edges of the tubera ischii, one inch and six-eighths.

Ditto, posterior edges of the tubera ischii, two inches and four-eighths.

From the right spine of ischium to the coccyx, six-eighths of an inch.

Ditto, left spine of ischium to the coccyx, five-eighths of an inch.

Depth of the fore part of the pelvis, three inches and five-eighths.

Ditto of lateral part of the pelvis, three inches and six-eighths.

Ditto of posterior part of the pelvis, three inches.

Distance, externally, between the bottoms of the acetabula, one inch and seven-eighths.

Ditto, internal measurement, one inch and three-eighths.

Ditto between the superior spinous processess of the ilii, seven inches and a half.

Ditto between the inferior spinous processes of the ilii, fou inches and seven-eighths.

Ditto between the centre of the crests of the ilii, ten inches.

Ditto from the upper edge of the fourth lumbar vertebra to the inferior spinous process of right ilium, one inch and six-eighths.

Ditto, of left ilium, two inches and six-eighths (see sketch, fig. 3).

Remarks.—Mollities ossium, under which this woman suffered, usually commences during pregnancy; but in her case it most likely began at the time she first complained of pains in the back and hips. The disease now most probably made little progress until her second pregnancy, when no doubt it was much aggravated; but its ravages on the pelvic bones were chiefly made during her last gestation.

Some of her relations had suffered from rheumatism, and it was stated that she had laboured under the same complaint; but for anything I know to the contrary, they (her relations) might have had the disease now under consideration. The predisposition, then (most likely hereditary), remained com-

L

paratively undisturbed until roused into activity by those changes which are produced on the vascular and absorbent system by impregnation. Her occupation as a weaver no doubt greatly contributed to increase the degree of pelvic distortion.

The projection of the promontory of the sacrum and lumbar vertebræ mechanically impeded the passage of the fæces into the rectum—which in such cases is always an additional cause of constipation—and which almost invariably occur to a greater or less degree in ordinary pregnancies.

The enema apparatus should have a long flexible tube (like that used to throw fluids into the stomach), so that its point can be carried beyond the altered pelvic brim.

An ordinary female catheter is too short to answer its purpose in cases of this sort; it should be equal in length to that of the male. By the deflected state of the uterus, the bladder is forced forwards and downwards, whereby its cervix is lengthened and compressed upon the pubes. This altered position of the bladder requires a different method to pass the instrument. The point being introduced into the meatus (which lies in the upper part of the anterior pelvic slit, and a little behind the lower part of the symphysis pubis) the hand part must then be depressed and carried backwards towards the extremity of the coccyx, and the instrument must afterwards be cautiously moved on into the bladder.

The stethoscope affords us information of two kinds—positive and negative; in its application here we only derived the latter kind. The pulsation of the infant's heart could not be heard, and from the absence of its sound, coupled with the odour of the vaginal discharge, we concluded it was dead. The functions of the placenta had long ceased, and therefore no *soufflet placentaire* was audible, so as to point out to us the location of this organ. If it had been otherwise I might, and indeed ought, as far as possible, to have made the incision into the uterus so as to avoid cutting on and into the placenta. The importance of such a procedure I had stated before the operation to Mr. Dunn, and requested that he would, as expeditiously as possible, extract the infant, and at the same time pass one hand to guard its neck from spasmodic seizure by the uterus, if perchance I should unfortunately cut upon the placenta.

I have elsewhere (*Edinburgh Medical and Surgical Journal,* vols. lx. and lxvii.) ventured to state this (the incision, and consequent partial separation of the placenta) as a cause of this irregular contraction of the uterus. If the child had been alive, most likely such a grasp would have destroyed it before it could have been extricated.

It is to be lamented that the existence of labour had not been earlier known ; perhaps a different termination would have taken place. To wait for the changes which usually occur in the os uteri during the first stage of normal labour is a great mistake in those cases in which there exists great pelvic distortion.

This part of the uterus is not often to be felt, and when it is accessible it is found a little open ; its full dilatation having been prevented by the vicious conformation of the brim of the pelvis. The evil arising from this procrastination ought to be a beacon to us in future, and guard us against vainly waiting for such a change.

The highly inflamed state of the bowels, which existed before the operation, and their pathological condition after death, together with that of the os and cervix uteri, &c., afforded indisputable evidence of the mischief which had been inflicted during her protracted labour.

The term protraction should be relatively considered. A woman (as in this case) in a bad state of health cannot safely endure the same degree of pain, or for the same length of time, as one who is well ; nor will the os and cervix uteri bear pressure very long without structural injury. Surely there is here sufficient proof of the necessity of the Cæsarean section : for neither the presentation of the infant, nor the os uteri, could be touched. Independent of these circumstances, I am bold to assert the delivery could not have been performed by the perforator and crochet, even if assisted by the marvellous power of the osteotomist, provided the implements were used by the most experienced and celebrated anti-Cæsareanist. No practitioner then, however warped his opinions are, will presume to attribute the death of this poor creature to the operation ; but, on the contrary, will place the case in the catalogue (already too great) of those in which the death has happened in consequence of having too long deferred its performance.

CASE (Tab.) 42 (*Unsuccessful*).—I am enabled to lay before the profession the following interesting case, through the kindness of the brother of the late Dr. Hardy, who had promised to give it me previous to his death :—*

"Betty Wilcock, the subject of the following operation, was within a few weeks of completing her forty-ninth year, as appears from the baptismal register, dated October 26, 1776. Her general health was good until within the last seven years, during which she experienced a dull heavy pain in the back and hips, and felt herself gradually growing weaker. Five years ago she consulted me respecting the pain in the loins; she was advised to take laxatives, with mercurial alteratives, and afterwards to try the sea air, from which she derived considerable benefit. Her health was best during the summer months, the pain and weakness always increasing on the approach of winter; she decreased several inches in stature; and walked with considerable difficulty, and for the last two months of her pregnancy she was carried to and from bed. She was the mother of ten children, the youngest of whom was six years of age, and I was informed her labour at that time was more lingering and protracted than any she had previously experienced. I had not seen her from the time she consulted me about the pain in her loins until I was requested to attend her in labour. The pains commenced on Tuesday night (June 25, 1825), and, continuing, she sent for me about twelve o'clock on Wednesday night. Being engaged, my senior apprentice went and remained with her until Thursday noon, when I visited her; I found that the membranes had ruptured spontaneously, and the water dribbled away; the pains came on regularly every six or eight minutes. On making an ordinary examination I could not reach the child, and, on trying to introduce my hand, experienced great difficulty. The tubera ischii approached each other within an inch and a half; the angle formed by the ossa pubis was very acute, and from the tuber ischii to the symphysis might be about three inches. On carrying my hand flat between the rami of the pubes I could introduce it above six inches, and

---

* The patient, Betty Wilcock, belonged to Mr. James Bailey, surgeon, Blackburn, who operated, and drew up the case as here detailed, and lent the manuscript to Mr. Hardy. Attested by his son.

found the child's head presenting, with a portion of the funis. The superior aperture was very small; the junction of the two last lumbar vertebræ, with the os sacrum, projected into the cavity of the pelvis, so that the antero-posterior diameter did not appear more than one inch and a quarter. Over this projection, and on the symphysis pubis, the head of the child lay. The pulsation in the funis was very strong. Though she had been in labour for more than forty hours, she had no fever. Pulse, 84; tongue coated with a brown crust, but quite moist. At eleven o'clock, finding the labour did not proceed, and her strength declining, I sent for Mr. Hardy, of Whalley (a gentleman who had extensively practised the different branches of the profession for upwards of thirty years). He arrived about one o'clock on Friday morning, and on a very careful examination found everything as above detailed, and agreed with me on the impossibility of delivering the woman with the crochet. On explaining to her the circumstances of her case, and the necessity of having recourse to the Cæsarean section, as the only means of affording her relief, she willingly acceded to it. At three o'clock A.M., Friday, I sent for Dr. Martland, of Blackburn, who carefully examined the patient, and fully agreed with Mr. Hardy and myself on the propriety of performing the Cæsarean operation. The bowels having been freely moved by an injection, and the bladder emptied, and the necessary preparation being made, at seven A.M. the patient was placed on the table. The abdomen was very pendulous, and rested upon her thighs. Dr. Martland placed his hands on each side, and firmly supported it. I commenced the incision through the integuments, beginning about an inch and a half below the umbilicus, and extended it down the linea alba, to within two inches of the symphysis pubis, making it six inches in length. The parietes of the abdomen were very thin, and the uterus was seen rolling through the peritoneum, which being divided by the probe-pointed bistoury, the uterus was fully exposed, and appeared of a light rosy colour. I commenced the incision into the uterus a little above the cervix, and carried it for the space of five inches towards the fundus, being as cautious as possible in avoiding the large sinuses. The cut surface of the uterus was fully half an inch thick. A large bag of water, of a pale whey colour, imme-

diately protruded, which, on being opened, discharged about
sixteen ounces. The left arm of a child now presented, and
it was easily extracted, and soon was very lively. Very soon
after the delivery of this, the right leg of another made its
appearance, covered by the membrane, through which a
quantity of meconium was seen. This child was also delivered, and shortly after birth exhibited signs of animation.
This was the child that presented its head and funis over the
superior aperture of the pelvis. The apex of the head was
corrugated and contracted into a small cone, its circumference
scarcely exceeding two inches and a quarter. A very moderate
hæmorrhage succeeded the extraction of the placenta. The
divided edges of the uterus, which now appeared above an inch
in thickness, were brought together and kept in close apposition for a short time; and so great was the contraction that,
in a few seconds, the wound did not exceed half its former
size. Through the firm and judicious support of the abdomen
and uterus none of the intestines were seen; the former was
closed by the Glover's suture; long slips of adhesive plaister
were applied, and a many-tailed bandage, made of flannel, was
firmly bound round the body. During the operation the
patient was rather faint, but a little weak brandy and water
revived her, after which she was removed into bed. The pulse
was 108, and tolerably full; there was a plentiful discharge of
lochia, and she expressed herself very comfortable. I visited
her with Mr. Hardy about half-past eleven A.M., and found she
had been quite easy since the operation. Her tongue was
moist, but coated with a light-brown crust; pulse, 104. She
had no pain or tenderness in the abdomen, had a very plentiful
discharge of lochia, but she experienced slight pains in the
uterus, resembling after-pains. I ordered her to take the
simple saline effervescing drafts every two hours.

Half-past seven P.M.—I visited her with Dr. Martland.
Pulse, 120, tolerably strong and regular; tongue rather dry,
but not furred. She had comfortable perspiration during the
day; complained of a twitching extending from the epigastric
region to the right shoulder. She had passed urine since the
operation. There was no distension of the abdomen, nor any
particular pain on pressure; a plentiful discharge of lochia. I
introduced the catheter and extracted sixteen and a half ounces
of urine, after which she felt relieved.

R Sol. acet. morph. ♏viij ; aquæ puræ ʒj. M. Ft. haust. h. s. s.

R Hydr. sub. gr. 4 ; pulv. ipecac. co. Ɔj ; pot. nitr. ʒiss. M. et div. in pulveres viij ; quorum sumat i sextis horis. Cont. mist. sol. efferv. ut antea.

July 2nd, seven A.M.—In the early part of the night she had a great deal of pain in her bowels, constant, but not resembling after-pains. The uneasiness about the stomach had left her, and she slept at intervals through the night, and continued to perspire gently. About twelve o'clock she vomited a little sour water, by which she was relieved. Lochia plentiful. Catheter introduced, and about ten ounces of urine drawn off. Pulse 120, strong and regular; tongue moist and coated with a slight brownish fur. Through the neglect of her attendant the draught was given in divided doses, and the powders were omitted. Inspiration 26 per minute; feels quite comfortable; abdomen free from pain.

R Magn. sulph. ʒiij ; mag. carb. Ɖj ; inf. senn. ʒiiss. M. Ft. mist. cujus sumt. coch. ij, quâque horâ donec alv. resp. Cont. mist. sal. efferves.

Seven P.M.—She had been tolerably easy since morning, but had experienced slight twitching pains, extending from the stomach to the uterus. Pulse 120, rather full; tongue moist, but coated with a little brownish crust. She had a slight cough, which gave her much uneasiness in the left hypogastric region. Lochia copious. Abdomen rather more painful on pressure, and more distended, evidently with flatus. Catheter introduced; only six ounces of urine were discharged. The opening medicine was not given according to direction, and the powders were again omitted. An enema was administered, which moved her bowels freely, and gave relief.

July 3rd.—I was called to her about half-past one A.M. She had been seized with sickness, and ejected some brown water from her stomach, which came up in mouthfuls, and caused her much pain in the epigastric region. She was considerably relieved by bleeding. The tenderness in the abdomen abated, and pulse 120, more feeble. The blood had a very thick inflammatory coat, and was cupped. To take five drops of sol. morph. each hour. Emplast. lyttæ reg. epigast. applic.

Eleven A.M.—Sickness left her soon after taking the drops.

She remained tolerably easy, and slumbered.   Had a few of the twitchings from the stomach to the uterus.   Not much tenderness felt on making pressure, but abdomen more distended, and, on being struck with the finger, had a tympanitic sound; pulse 130, feeble and intermitting; countenance lost its animation; catheter introduced, and about five ounces of urine extracted; bowels not moved.

℞ Hydr. submur. gr. vj; extr. col. c. ℥ ss.   M.   Pil. vj.
Cap. j. sec. quâque horâ.

Five P.M.—Soon after I left her the regurgitation of water from the stomach returned.   She became more restless; pulse fluttered, and was very feeble; her hands were covered with a cold clammy sweat, and she appeared to be fast sinking.

Seven P.M.—In a state of stupor, from which she could be roused, and would give sensible answers to any question, but soon relapsed.

She expired about half-past eight—sixty-one and a half hours after the operation.   The children feed very well, and are in perfect health.

Remarks.—The foregoing case is unique—at least in Great Britain and Ireland—in the annals of medical literature.   It affords an example of the preservation of the lives of two infants; and, if the operation had been earlier performed, and the after-treatment strictly adapted to the exigences of the case, most probably the mother might also have been saved. These twins were girls, and most likely arrived at womanhood; as it is stated, in an appended paper to details of the case, they presented themselves for confirmation after they had passed their sixteenth year.

Betty Wilcock had borne ten children, and, in all but one, her labours were natural and easy; this, the last, was protracted; so that it may be supposed the pelvis had at this period suffered in its conformation.   The disease which caused the pelvic deformity was malacosteon, which evidently at first slowly advanced, mainly owing to the long interval (six years) between her last (tenth) and this (her eleventh) labour.   Gestation invariably aggravates this disease; hence the increased relative difficulties experienced in successive labours in cases in which it exists.

There is no positive pathological evidence what caused her

death, as no post-mortem examination was made; but from the symptoms and treatment we may suspect that inflammation of the peritoneum, or some of the abdominal viscera, had taken place.

The duration of her labour (eighty-three hours) is quite sufficient to produce all the mischief which ensued; and the result affords another striking example of the danger of procrastinating the operation. The suture (Glover's) used is not suitable to close the abdominal wound during life; and though unwilling to make any observation which tends in the least degree to detract from the character of the operator, yet it is a duty to warn others against a practice fraught with danger: it must decidedly tend to injure the peritoneum.

CASE (Tab.) 49 (*Successful*).—At half-past one o'clock A.M., Thursday, November 20, 1845, I visited Mrs. Sankey, Bedford Street, Salford, who was then in labour. I met Mr. John Goodman, from whom I learnt the case was one of extreme distortion of the pelvis, &c. She first felt slight pains about eleven o'clock on Wednesday morning. Sometimes they were stronger, and continued to vary in severity until about three hours before my arrival, during which time they were uniformly stronger. The membranes ruptured some time in the afternoon, and there was a slight dribbling discharge at intervals up to the time of my first visit. The tongue was clean and moist; her countenance was calm and composed; her bowels had been freely evacuated; and she had regularly discharged her urine; she was free from sickness and vomiting; pulse 84 at first, but temporarily rose to 100. Placing my hand upon the abdomen, I found the uterus of a spheroidal shape, but less so than I had observed in former cases: it felt firm and free from pain when I pressed upon it. The relative position of this organ was now much altered: it hung over, and rested on the pubes; and the fundus projected so much forward as to actually form its anterior surface. Connected with, and in reference to, the axis of the vagina, it presented the retort form. By a vaginal examination, I found the outlet of the pelvis considerably diminished by the close approximation of the rami of the ischii and pubes, which, jutting forward, had nearly destroyed the arch, and leaving in its place only a narrow slit which would barely admit the point of the finger. The tubera ischia were so near as scarcely to allow two fingers to lie in the transverse diameter; the antero-posterior was also much shortened by the great incurvation of the coccyx and lower portion of the sacrum. The cavity was diminished by the several bones which unite in each acetabulum being forced inwards, backwards, and upwards. It was difficult to measure the brim; its figure was tripartite.

This alteration in shape was produced by the falling down of the promontory of the sacrum and lumbar vertebræ, and by the ischia and pubes being pushed backwards and inwards, and the jutting forwards of the symphysis and rami of the pubes. The antero-posterior diameter was not more than an inch and a half at its widest part on the right side, and in some parts not an inch. The anterior lobe of the opening was not more than would admit the finger edgeways (see sketch, fig. 4). The vagina was moist, and had only its natural temperature; it was free from swelling. The os uteri, or the presentation of the infant, could not be felt.*

On returning down stairs to Mr. Goodman, I cordially agreed with him that the Cæsarean section was the only means by which delivery could be performed. I suggested the propriety of having the stethoscope applied to testify whether the infant was alive, as the mother affirmed, and also to ascertain the location of the placenta. At this stage I also requested that Mr. Winterbottom, surgeon, should be invited. A messenger (Mr. S.) was dispatched for Mr. Winterbottom, who was desired to bring with him a stethoscope. I now remarked on the danger of cutting into and partially separating the placenta during the operation. Auscultation indicated that this organ was attached to the anterior and right lateral portion of the uterus, and corroborated the statements made by the patient that the infant was alive.

I now raised the fundus uteri, and Mr. Goodman made an incision, from seven to eight inches long, a little to the left of the umbilicus, and divided the integuments, which were extremely thin. A corresponding opening was now made through the uterine tissue, which was about a quarter or one-third of an inch in thickness, just bordering on the edge of

* See Fig. 4th.

the placenta. I now passed my hand and seized one leg of the infant, and extracted it; and when it passed to the head, Mr. Goodman assisted its escape through the wound, and afterwards expeditiously extracted the placenta entire. The infant—a girl —was alive. There was very little blood indeed lost, and the uterus contracted and lowered itself to its resting-place. The intestines and omentum protruded during the operation, but were then restrained, and afterwards replaced and retained in their natural position during the time the ligatures were introduced. I swept my finger round the boundary of the uterus, and through the wound in its cavity, to ascertain if any portion of the viscera had got into it. The interrupted suture was used, and the thin integuments further brought together by strips of adhesive plaister; and over all a bandage, just tight enough to give support, was applied. There was a calmness and resignation of mind throughout the whole operation which indicated an enduring Christian spirit. No complaint was made, nor murmur heard; and she expressed herself as happily relieved.

The pulse now beat 92. She had a little gulping; the skin felt warm. Habeat tr. opii ʒiss; in one or two hours pulse 94; slight retching; and a very trifling oozing of blood was seen on the bandage; her lips were naturally red; respiration 26. Fresh bandage and a compress applied. At a quarter past nine o'clock A.M. tongue clean; countenance natural; respiration 24; about two or three ounces of blood lost; a negative plan recommended—mist. acac. ʒj, vel. ʒiss—ter in die.

Eight o'clock vespere.—Pulse reduced from 100 to 90; countenance good; had complained of tightness of the bandage; the belly a little tumefied, but not tender; there is no bleeding. To take ext. hyoscyam. gr. x.

November 21st, ten o'clock A.M.—Had vomited a sour fluid during the night. Pulse 86; countenance natural; tongue clean and smooth; skin cool; lochial discharge natural; bowels not moved; had passed her urine. Ordered an enema as follows:—Spt. terebinth. ʒij; ol. ricini ʒj; vitelli ovi q. s. et aq. hord.

Half-past one o'clock P.M.—Tongue clean; she has vomited a dark-coloured thick and sour fluid, which in appearance

looks like fæces.   Enema has not operated; ordered to be
repeated ; to take barley water.

Nine o'clock vespere.—Has had no stool; vomiting continues ; belly tympanitic.   The compress being removed, the
site of the lower portion of the abdominal wound was tumid,
and on raising the plaster the wound itself was seen gaping,
and a fold or convolution of intestine had protruded between
(in the interstices of) the two ligatures.   The bandage and
plaister were now removed.   An attempt was made to reduce
the bowel, and the edges of the wound well adapted, and fresh
strips of plaister and a three-tailed bandage applied.   Ordered,
at her request, ol. ricini f3ss : to be repeated if necessary.

22nd, nine o'clock A.M.—Had some sleep; pulse 100; had
vomited the oil, and also some of the same fluid; tongue
clean ; belly less swelled, but slightly sore; there is a considerable bloody serous discharge from the wound; has passed
urine ; lochial discharge; has had no stool.   To take calomel
gr. v ; a quantity of gruel to be thrown up by the syringe
with the œsophagus tube passed high up towards the sigmoid
flexure of the colon.

Seven o'clock vespere.—Has slept several times; pulse 90
to 95 ; feels better; skin cool; vomiting ceased ; belly less
swelled, and easy; another enema had been administered, but
it procured no stool.   Ordered calomel gr. v ; sacch. alb. gr. vss.

℞ Magnes. sulph. 3vj ; magnes. calcinat. 3ij ; tr. cardam.
            comp. 3j ; aq. cinnam. 3iij.   Coch. larg. j tertiis horis.
            Repr. dos. anod. noct.

23rd, ten o'clock A.M.—She had slept; pulse 90 ; tongue
clean ; had vomited only once a light-coloured fluid; bowels
opened four times; skin warm and perspiring; very slight
abdominal uneasiness.

Five o'clock vespere.—Had visited her before, and found
her going on well; a considerable serous discharge from the
wound ; had not vomited ; has had three stools ; passed urine ;
has had several sleeps.   To continue mist. acac. &c.

24th, ten o'clock A.M.—Pulse 95 ; tongue clean, but redder
in the centre; has very little abdominal pain on pressure;
bowels purged two or three times ; has passed urine ; says she
feels weak.   Pergat.   To take beef-tea.

Half-past seven o'clock vespere.—Pulse 100 ; tongue clean

but sore; belly less tumefied, and bears pressure; a considerable bloody serous discharge from the wound; had been twice purged, attended with griping pain. To continue the medicine; also to take half an ounce of the following mixture if the diarrhœa continues, and also to take one pill :—

R Cretæ ʒj; mist. acac. ʒiss; aq. dest. ʒj; tinct. opii ʒj. M. ft.

R Opii gr. ss; conf. arom. q. s. Fiat pil. h. s. s.

November 25th, ten o'clock A.M.—Pulse 94; tongue clean; had slept; lochial discharge natural; had passed urine; bowels not again moved; gentle perspiration; belly easy. Pergat. Ordered chicken broth.

Half-past four P.M.—Pulse 85; symptoms all favourable; has a cough; taken freely of chicken broth and beef-tea. Ext. hyoscyam. gr. x h. s. R Mist. acaciæ ʒij; tr. camph. co. ʒiij; syrup. rhœad. ʒss. M. fiat tinct. cujus cap. cochl. min.j sub urgent. tuss.

26th, ten o'clock A.M.—Pulse 95; tongue clean; skin a little hotter; bowels unmoved and tympanitic; had passed urine; her countenance a little more anxious. The bandage and plaister were removed; the ligatures were detached from their hold; the integuments expanded to such a degree as to present a large surface of the bowels, which were uniformly covered with lymph; the incised edges were firmly agglutinated to the contiguous parts; attempts were made to bring the integumental edges nearer together, and the bowels were gently pushed back, and broad straps of adhesive plaister were tightly applied, and a six-tailed bandage, so as to afford firm and equal support. Ordered Mist. magnes. sulph. ʒvj; magnes. calcin. ʒij; aq. ʒvj; cujus cap. ʒj tertiis horis donec alv. resp.; and the following form of enema to be administered if necessary :—

R Sp. terebinth. ʒij; tinct. assafœt. ʒij; aq. ʒvss : to be added to ʒviij of barley water.

Six o'clock vespere.—Pulse 120; skin hotter; has great depression; tongue dry; countenance anxious; bowels unmoved; complains of pain in the belly; had passed water. Ordered the enema to be immediately thrown up. She has just vomited. Pergat. An anodyne to be prescribed.

27th.—Took morph. acet. gr. ⅛ cum ext. pap. gr. iij. She

has had four motions; has slept well, and generally better. Pulse 90 to 96. Cont. mist. acac. tinct.; beef-tea; and take anodyne at bed-time.

28th.—Pulse 96; has slept well; has had two motions; a considerable discharge from the wound; all the other symptoms better except the cough, which is rather more troublesome. Continue medicine and diet—and take isinglass.

29th.—Pulse 84 to 90; slept well; bowels twice naturally moved; all the symptoms more favourable; a considerable discharge on the bandage. Upon removing the bandage and dressings, the wound presented a healthy granulating surface, and covered with pure pus. There was a small slough at the lower edge. The integuments below, in the bend of the thigh, were irritated and excoriated by the discharge. Continue diet and medicines.

30th.—Pulse 86 to 100; tongue clean and smooth; bowels moved. Complains of sore throat, which is red, and slightly aphthous. Cough very troublesome, and, from its force, produces a sensation of forcing some part through the wound, and, on examination, a portion of bowel was found protruded. The lower part of the bandage and dressing being now removed, the wound was seen not to be so well; the granulations were paler; the discharge from the left was blackish from the slough. Fresh pledget of lint and plaister applied. Continue the same medicine and diet, and also to have eggs, and the following mixture:—

R Sodæ sub-borat. pulv. bene ℨiij; syrup. rhœad. ℥ss; mist. acaciæ ℥ijss. M. cujus capiat paululum subinde.

R Morph. acet. gr. ¼; ext. pap. alb. gr. iij. Fiat pil. h. s. s. et rep. in horis 2 si op. sit.

December 1st.—Pulse 100; all other general symptoms better. A considerable discharge from the sore, which was dark-coloured and offensive; a protrusion of the granulated portion of the sore through the dressings, which were now removed. The aspect of sore pale; its left edge was sloughy in places. The dressings were changed, and pledgets of lint with cerate over. Double folds of linen were placed, and a six-tailed bandage, to support as far as possible. Our object was to form a barrier to further protrusion, as, from the intimate adhesion which now existed among the convolutions of the bowels themselves, and between them and the integuments, it would be mischievous to attempt to do more than barely to defend and to support these parts. On the right side the old and the new surface presented one plane. To take a generous diet, and porter; and also the anodyne pill.

2nd.—All the general symptoms favourable. Had first refused to take porter, but afterwards did so. Wound looks better; granulations redder; pus laudable; cicatrization on left side proceeding; sloughing appearance on the right side much better. Continue diet, porter, and a night-pill. At the evening visit she reported she had taken a beefsteak at dinner, and porter; and we found her eating the leg of a fowl, and again drinking porter. Cautions given not to get into excess.

3rd.—Mr. Goodman, at a later visit last night, found her pulse considerably more frequent. She complained of the wound. On examination he discovered that a large loop of intestine had protruded on the left side, which he unsuccessfully attempted to reduce. He placed a pledget of lint with cerate over it. This morning the pulse 120; bowels not moved; she had vomited some undigested animal food. The body linen was very wet by discharge, which had a pungent disagreeable smell; the dressings adhered so firmly that they were with difficulty removed. A loop of intestine, of a horseshoe shape, covered with lymph, and fixed by adhesion; it was very much distended. Cautions but unsuccessful attempts were made to reduce it. I suggested the propriety to puncture it with a needle; a sharp-pointed bistoury was, however, used, but nothing but a little blood escaped. Another unsuccessful attempt was made to reduce it. Broad slips of plaister and a bandage applied. An enema, with oil, &c., to be thrown up with the syringe and œsophagus tube. To discontinue the porter and animal food; to take chicken broth. Ordered some coriander or mustard seed to be taken, and the anodyne pills.

4th.—Bowels freely moved several times, but the seeds were not seen; all the symptoms better. Continue.

5th.—Bowels three times moved, and in the last motion the coriander seeds were seen; every other general symptom favourable. The wound healthy; the union on the right side complete, and the cicatrization proceeding : on the left side the loop of the bowel still lay prominent, and the edges of the sore were everted, and in an unfavourable position for this reparative process. Continue diet and anodyne pills.

6th.—Pulse 96 ; bowels unmoved; slept well; wound looking healthy, and lessening in diameter; cicatrization rapidly proceeding : the surface over the protruded intestine loop granulating ; this portion of the bowel immovably fixed down. Continue medicine and diet; to have an enema and her anodyne pills.

7th.—Pulse 90 ; bowels twice very freely moved, and in one motion some coriander seeds were found ; a profuse irritating discharge through the dressing. On exposing the parts, an opening was found at the lower end of the protruded bowel, through which fæces had passed, and were now escaping. The enema had not been given, but it was now directed to be administered. To continue to take the anodyne.

8th.—Pulse 90 ; slept well; bowels once opened by the enema. The size of the wound is considerably diminished, and its surface looking healthy. Fæces in great abundance were discharged through the intestinal opening. To continue the same plan. Ordered an enema and anodyne pills.

9th.—Pulse 84; bowels opened by enema ; the wound is diminished in size, and looks healthy. The fæces are largely discharged through the opening in the bowel. To continue ; to try to eat some calf's foot; to have an enema and the anodyne.

10th.—Pulse 80 to 84 ; the wound considerably lessened, and its entire surface looks well. The apparent size of the protruded bowel is greatly diminished, and is covered with healthy granulations ; its horse-shoe character is nearly lost. Great quantity of feculent matter comes through the opening; portion of undigested calf's foot was observed in the discharge. Ordered a little wine. Continue diet ; to have an enema and pill.

11th.—Pulse 74 ; wound lessened, and looks well; bowels moved by enema; fæces in great abundance from the opening. To continue wine, &c.

12th.—Pulse 72 ; had two stools, which were too light in colour, which perhaps depends upon the escape of the biliary fluid too soon, as the fæcal discharge by the opening has been all along very deeply tinged, and is still in great quantity. The sore is now considerably lessened; it does not measure more than four inches in length, and two to two and a half inches in breadth. The protruded intestine so far retired as not to be seen, except when the integument is drawn aside. The integumental edges of the sore can now be approximated. To continue the wine, &c.; to have an enema; to take a pill.

13th.—Pulse 74; right edge of the wound was more inverted, and which had been produced by approximating the two sides; bowels slightly moved; considerable fæcal discharge from the wound, and in it some coriander seeds were seen, which had been taken early in the day. The wound dressed in its longitudinal direction. Continue the wine, &c.; to have an enema; to take the pills.

14th.—Pulse 72; has had no motion; granulations paler; the discharge of fæcal matter greater through the wound; a sponge compress was applied over the opening, which was to be removed at intervals; it was, however, permanently discontinued in the evening. Continue the wine. To have jelly, blancmange, isinglass, and to be allowed a partridge.

15th.—Pulse 76; small hard fragment of stool brought with the enema; countenance more dejected; complains this morning of a load at the stomach, which, in the course of the day, ended in vomiting of a quantity of undigested animal food: the fæcal discharge great from the wound; as the granulations were exuberant, the sore was dressed with dry lint. Wine increased; continue altera.

16th.—Pulse 74; wound less, and looks better; bowels not moved. To only have some broth, with a little of the chicken pounded in it. Continue medicine; to have an enema.

20th.—No particular change in the aspect of the case up to this date. Pulse now 74; the fæces still largely discharged from the intestinal opening; the protruded portion of bowel very red from the everted mucous membrane. Mr. G. passed his little finger, and carried this part in; a compress was applied, and supported by plaister.

31st.—During the interval, the pulse nearly uniformly 72. Bowels some days spontaneously moved once or twice; but, nevertheless, an enema was daily administered. All the

general symptoms continued to improve, and the excoriations
produced by the discharge were attended to by washing, &c.
The everted mucous membrane of the bowel which appeared
through the opening had at first rather increased, but now
was considerably less. Compresses or pads were applied, &c.
The discharge of fæcal matter through the opening had
gradually lessened.

January 10th, 1846.—As in the last report, the pulse had
uniformly continued steady. The bowels were sometimes moved
more than once a day; the enemas were regularly administered; the everted state of the mucous membrane of the
bowel varied in degree and colour. Her spirits were very low
on the 4th, in consequence of the very great discharge which
took place through the opening in the bowel. The treatment
consisted of small pads placed immediately over the aperture,
being supported by small slips of adhesive plaister, and then,
upon the integuments and over the other, a larger pad, &c.,
fixed by broad straps and a bandage.

17th.—Since the last report, the pulse continued natural,
and she slept well, and her appetite was good. The bowels were
naturally opened once or twice a day, and therefore the enema
had been discontinued until the 14th ult., when the bowels
being only partially moved, and as the belly had become more
tumid, they were again used. The discharge from the intestinal opening was at first less solid and more serous, but afterwards it became again more feculent. Compresses of different
kinds were applied, and one suggested by me made of layers of
caoutchouc was used, but did not succeed. On this day the
fæcal discharge was much greater, which I accounted for by
the opening, which was about the size of a silver fourpenny-
piece, having changed its relative position with the integuments. I recommended that a conical piece of sponge should
be placed with its point inwards, and supported by plaister and
bandage. Mr. Goodman feared that it would be injurious to
use the sponge, as it might dilate the opening and induce
ulceration. My reply was, that I did not fear the first effect;
and if the second happened, it would only be limited in
degree, and advantageous. I thought at this time the edges
of the wound were in a favourable position to be pared, and a
ligature used to bring them into apposition.

22nd.—Pulse kept natural, and the bowels were daily moved twice. The sponge had become a little swelled, and the edges of the opening bled a little. It was re-applied on the 20th. The bandage was very wet, and was attended by a disagreeable ammoniacal smell, which no doubt arose from the decomposition of the discharged fluid, which had taken place in consequence of the parts not having been dressed the day before. The surrounding integuments were considerably irritated and excoriated. At my desire, a simple dressing with lint and cerate was adopted, and a very loose bandage applied. The patient sat up, but the horizontal position was advised. On this day (22nd) the eversion of the mucous membrane was greater, and the opening had slightly increased. A piece of sponge was recommended by me, but a compress of lint was applied at Mr. Goodman's desire.

23rd.—Pulse good ; bowels twice moved ; the integumental excoriation rather extended ; the intestinal opening is a little larger. Mr. Goodman proposed touching its edges with argent. nitrat. ; but I considered that the surface would heal underneath before the eschar separated, and suggested the potass. fusa. Mr. Goodman thought it would be difficult to limit its application, although my opinion was that its action might be controlled. I recommended instead of it, the edges of the opening to be pared off, which Mr. Goodman did by a lancet and bistoury. This operation was not so effectually done, in consequence of the tender and fragile state of the part. The edges were approximated, and supported by slips of adhesive plaister. The integuments being drawn together, formed a covering and compress to the aperture.

31st.—During this interval the pulse was natural, and the bowels some days moved spontaneously, and always responded to the use of an enema. The intestinal opening remained much the same ; and the discharge from it, although not equal in amount every day, yet it was sometimes very great. Various methods were tried to control it ; pitch plaisters in different ways, and compresses, were used at the suggestion of Mr. Goodman, but they invariably failed to do good ; on the contrary, mischief was done by them. At my request, pads made from caoutchouc, of different sizes, and placed differently, were employed, but equally unsuccessfully. I recommended broad straps of plaister from the infirmary to be tried, which also failed. As considerable irritation and excoriation had been produced by the pitch plaister—indeed, the pain and inconvenience sustained by the patient was such that she not only requested a discontinuance, but strongly protested against its use—my opinion was unfavourable to their use, as I considered disadvantages from these arose, not only on account of the irritation, &c., arising from the pitch itself, but from the mechanical effects, by inflecting the integuments upon each other, and also by impeding, by the great pressure produced, the peristaltic motion of the bowels. On the 30th, I saw the patient alone, and thought it would be better to apply a simple dressing.

February 1st to 13th.—It appears that a great discharge issued from the opening after the last dressing; so that the nurse, at the desire of Mrs. Sankey, had removed them, and brought the edges together by adhesive plaister. Long strips of plaister were again recommended by Mr. G., and very tightly applied, some of which were passed round from one side of the spine to the other.

Great integumental irritation and excoriation ensued, and it was therefore thought desirable not to bind the parts so tightly. The edges of the opening were touched from time to time with the argent. nitrat. On Februray 2nd, I suggested the propriety of having a bandage made of some elastic but firm material, so that the diaphragm and abdominal muscles could move in opposite directions without the least impediment, and at the same time afford uniform support.

11th.—Mrs. Sankey stated that they had removed the quill, as nothing had passed through it. Mr. Goodman had placed the barrel of a quill as a conductor for the fæcal discharge; but this, it appears, did not answer, nor was it likely to do so, as it would be quite impossible to place such a body in a parallel direction with the course of the intestinal tube; and we cannot for a moment suppose that it could act by suction.

12th.—I found that pitch had been used again in dressing the part the day before. The evils were increased by this plan: the surrounding irritation and excoriation were seen to be much greater. Mr. Goodman ordered a new bandage, with gussets for the hips; and buckles and straps were attached to it, so that it could be tightened and fixed. This contrivance proved of great advantage to her. She wore it a considerable time, and Mr. Goodman states he had the satisfaction to find a progressive diminution of the fæcal discharge, which passed through the intestinal aperture.

Remarks.—Mrs. Sankey's occupation was merely domestic ; she was a slender, delicate, and spare-looking woman. Her complexion was pale ; her skin was swarthy ; her eyes were grey ; and she was of a leucophlegmatic constitution. She was naturally of a mild and placid disposition ; and her mind was placable and strongly imbued with sound, moral, and religious principles, which wonderfully supported her in her severe trial, the intensity of which it is scarcely possible for imagination to conceive. What can be more dreadful than the anxieties of a woman in the pangs of labour without hope of delivery ? She bore the operation with great fortitude, and scarcely made a moan ; and afterwards endured a tedious recovery with great patience and resignation.

Her age was forty-one years. This was her seventh pregnancy. Her first four labours were natural and easy : in the first, she was delivered of a living girl ; and, in the other three, of boys—all born alive. Her fifth labour was natural, but it was rather more tedious than her former ones ; the infant (a boy) was born alive. At the end of two years she was again in labour of her sixth child, which was protracted by pelvic distortion ; and Mr. Goodman, with Mr. Slack's assistance, delivered her by means of the perforator and crotchet.

The disease (mollities ossium) under which she now suffered, no doubt commenced after her fifth labour, which proved a little tedious. It is, however, very probable that the mischief inflicted on the pelvis was only slight at this period : and, during the interval between it and her next pregnancy, most likely the disease remained stationary. Whilst pregnant of her sixth child it rapidly advanced, and committed great ravages in the pelvis. Her general health suffered and her strength failed ; her stature became obviously less. Mr. Goodman kindly informed me the thighs somewhat bowed, but afterwards recovered the usual shape. During her labour, Mr. Goodman found the pelvic diameter shortened ; the antero-posterior only measured about two inches.

After a consultation with Mr. Slack, he delivered her with the perforator and crotchet; it, however, rarely happens that the disease is so rapid in its progress, and does so much injury to the pelvis, as to require these murderous instruments before those which are compatible with the safety of the infant have been used in a former labour.

After her puerperal recovery, Mr. Goodman judiciously prescribed for her tonics and other appropriate remedies, and recommended her to the seaside for the benefit of bathing. Her health was greatly improved, and perhaps the progress of the mollities ossium was temporarily arrested. The disease again returned, and produced effects on the pelvis, during her pregnancy, which rendered necessary the Cæsarean section for her delivery.

Many circumstances existed in this case which favoured its propitious termination. Her calm and tranquil state of mind, and the high degree of moral courage she had, were very advantageous.

An early rupture of the membranes, in protracted labours, especially if the pains are strong, is always to be deplored; but, most fortunately, this did not happen here very long before the performance of the operation. The water was not suddenly and completely discharged at once, but dribbled away. The pains were also very feeble, until a very short time before my visit. The conjoined effects of these were doubtless favourable, and prevented injurious pressure on the soft parts. The bowels were freely evacuated during her labour. By the aid of the stethoscope the incision was made in a direction to avoid cutting into the placenta, which is very important.

When the placenta is cut upon or into, partial irregular contraction of the uterus happens, and through such an accident the infant may be lost, as I have already stated; but besides this, the tissue of the uterus is likely to suffer from the force required to extricate it (the infant) from the spasmodic grasp of this organ. Besides, there may be a greater risk of hæmorrhage when the placenta is prematurely disrupted or separated.

The operation was timely performed in this case. There happened several important contingent circumstances, which I

shall briefly advert to. The abdominal parietes were so attenuated as only to allow a very slender hold to be taken by
sutures, and were quite unequal to resist the pressure they
were subjected to. By the force of the cough, &c., this weak
structure was torn through, and the edges being thereby set at
liberty, retracted on each side to a considerable extent, and
exposed the bowels fully seven to eight inches in one direction
by three to five in the other. Notwithstanding this serious
accident, the constitutional disturbance was only trifling in
degree and of short duration.

The changes which successively took place on the exposed
surface of the bowels, were truly astonishing. A thin trans-
parent effusion of lymph was first seen laid over the peritoneal
coat of the intestines, which entered a little into the depres-
sions formed between their convolutions. This thin gelatinous
material gradually became thicker and more opaque—evidently
the second step towards cicatrization. During this wonderful
process—for its effects were truly wonderful—innumerable
vessels were seen, destined to carry on this great work of
reparation. As this went on, there was a progressive and
perceptible diminution in the size of the surface; and when
completed, there was a cicatrix to be seen very little different
from one which is formed after the healing of a wound whose
edges have been kept in juxtaposition.

Symptoms of straugulation appeared, for the relief of which
several expedients were adopted; the protruded portion of the
bowel was ineffectually punctured. Nature, however, stepped
in, and immediately the symptoms were relieved by the formation of an opening through which a great quantity of
fæculent matter and flatus was discharged. This aperture, as
is mentioned in the reports, continued throughout the case.
Although this disaster protracted the patient's recovery, there
is no doubt it had, at the time it happened, a most salutary
effect. During the whole period of the case the bowels were
strictly attended to; and there is no doubt their ready obe-
dience to our measures (chiefly enemata) considerably contributed to her recovery.

The reparative process which happened in this case, and the
decidedly conservative constitutional power which this woman
had, prove beyond dispute that the opinion expressed by

my friend Dr. West (*London Medical Gazette*, No. 1,210, February 1851, p. 245) is not infallibly true, and therefore ought not to influence us and deter us from the performance of the Cæsarean section. The same restorative power was observed in the other successful case, which is already before the profession.

As it was important to preserve the infant, I strenuously urged Mr. Sankey to obtain a wet-nurse, and no time was lost in obtaining one. The infant continued to thrive, and lived rather more than seven months. Her death was occasioned by an hydrocephaloïd disease—the consequence of exhaustion from diarrhœa.

Sequel.—Mrs. Sankey became again pregnant; but, as I was not consulted in her case, I can only give the account published by her medical attendant (see *British Record*, p. 14). It is stated, "information was received on the 25th of September that Mrs. Sankey was again pregnant: at this time the catamenial flow had ceased for two months, but there was no enlargement of the mammæ or change in the areola of the nipples; no morning sickness was experienced, and there existed no perceptible change in the desires of the stomach, or in the organs of sensation; still there was a progressive increase in the size of the abdomen, and a feeling on the part of the patient that she was decidedly pregnant.

"Having at length determined upon the course to be pursued, we directed, at first, drachm doses of secale cornutum. to be administered daily, and afterwards twenty grains of the same, at more frequent intervals. On the 28th of September we commenced the administration of the infus. sabinæ, in gradually increasing doses, beginning with six grains; this was continued until the 12th of October, when half-drachm doses were administered, combined with the same quantity of secale cornut. ter in die. These measures, with the pil. aloes c. myrrh as an aperient, formed the method of treatment until the 29th of October, at which time Mrs. Sankey, experiencing no change in any respect, entreated us to desist from any further attempt. In consequence of our inability to detect any symptom by which to determine that the desired action of the remedies employed had taken place, we abstained from the further administration of remedial agents, with the

exception of the pil. aloes c. myrrh, as an occasional aperient.
After this period our patient remained in tolerable health and
spirits, and continued as free from the occurrence of uterine
pains, weight, or unpleasant feeling as since the commencement of the treatment, until the morning of December 7th,
which was more than a full month after the discontinuance of
these measures. On this day, being summoned to attend, I
discovered that during the night Mrs. Sankey had aborted a
fœtus of about two months growth, at which both the patient
and myself were well pleased; and, with the exception of
some vomiting, she continued to progress favourably for two
or three days. The placenta, however, was delayed; and,
although no hæmorrhage of any moment occurred, anxiety
was experienced on this account; it was detected protruding
from the os uteri, from which it was impossible to remove it.
Ordered sec. corn. two drachms, aq. fervent. three ounces; ft.
infusum stat. sumendus; and, for the sickness, a saline mixture was ordered to be taken during effervescence. The secale
cornutum was repeated on the following day, but during the
interval many attempts were made, both by manipulation and
instruments, to remove the placenta, which was now lying
impacted in the brim of the pelvis. On the third day I was
enabled sufficiently to lay hold of it, so as, by very strained
exertion, between two fingers used as forceps with the assistance of pressure on the abdomen, to succeed in extracting it
entire. This desirable accomplishment produced considerable
satisfaction, for Mrs. Sankey was already beginning to suffer
from the fœtid and decomposing condition of the retained
placenta. Some febrile action was now observed in the
system, and even typhoïd symptoms were, in some measure,
anticipated; and, after the removal of the placenta, the patient
complained of slight tenderness in the region of the old
wound. The hæmorrhage was so slight that it merely satu-
rated three napkins; the vomiting increased, and a mustard
poultice was applied to the epigastrium. Other remedies were
also employed, but the patient gradually sunk, exhausted by
continual vomiting, and the shock of parturition. She died
on the 12th of December, and, on the evening of the following day, we made a post-mortem examination of the body."

" *Post-mortem Examination.*—On inspecting the body an

orifice, the size of a pin point, was discovered in the situation of the original wound, and the linen around it was moistened by about six drops of slightly coloured serous fluid. On opening the abdomen, a general glueing and matting together of the arch of the colon and omentum to the adjacent intestines (in an area of the extent of eight or nine inches) and to the cicatrized skin of the abdomen, was observed; which, as will be remembered, was developed from, and healed upon, the exposed peritoneal covering of these viscera. Much flatulent distension of the colon existed, and it was fully proved that no Cæsarean section could have been again performed.

"The agglutination of the parts, through which the incision must have penetrated, rendered the performance utterly impossible. It would have been necessary (as it was in simply opening the body after death) to have dissected the skin from the subjacent omentum; and the dissection must have been continued until the whole of the skin under this covering had been completely separated from its adhesions to the smaller intestines; and they, also, would have required separating from each other before the uterus could have been exposed. Fatal as the case had proved, we could not avoid a feeling of satisfaction that the measures adopted had been directed towards the induction of abortion, instead of reserving the mother for an operation, which would have proved fatal in the very hour of performance. The gall bladder and duodenum were distended with black bile, and the uterus was empty, and considerably congested at its fundus. The cicatrix of the original incision into the uterus was well defined, and there was no adhesion of the fundus to any adjoining viscera. There were no other decided marks of inflammatory action." The opening into the cavity of the pelvis, instead of presenting its proper oval form, had assumed a tripartite or trilobed character.*

It is further stated that "a perpendicular section of the pelvis, showing the projection of the sacrum, ossi ilii, and the cavity of the vagina, &c., was about three inches in its perpendicular axis."

"From the pubes to the margin of the ribs seven inches and three-quarters, to the point of the sternum only nine inches."

* See Fig. 5th.

" The pubis and conjoined ossa ilii" were " seen projecting inwards and backwards, and thus diminishing also the vaginal
cavity " at the outlet. " The tuberosities of the ischium
(ischia) joined at the centre. The anterior fissure between these
bones was only half an inch in diameter, the posterior opening
was laterally two inches, and antero-posteriorly two inches and
three-quarters diameter."

Some of the above quotations were given as explanatory of
three small figures; but as they are intended to point out
some measurements, I have inserted them.

The aforesaid remarks are solely quoted for the purpose of
making the interesting case of Mrs. Sankey complete, and not
to indicate an approval of the course adopted.

The agglutination and matting together of some of the
abdominal viscera which took place, is attributable to the
yielding of the edges of the wound and the protrusion of part
of the intestines.

The adherent state of the abdominal viscera is amply sufficient to arrest gestation and consequently lead to abortion.
Therefore there is little doubt if the case had been left to
Nature the ovum would have been most assuredly expelled
without artificial interference.

Case (Tab.) 51 (*Unsuccessful*).—The following case has been
most kindly and liberally given to me by James Braid, Esq.,
M.R.C.S.Ed., C.M.W.S., &c., for the purpose of publication :—

" My DEAR SIR,—I beg to hand you the following brief
particulars of the case of Cæsarean section in which I operated
in 1847. So far as it goes, it fully justifies the doctrine you
so ably and properly advocate :—

" About one o'clock P.M., on the 15th of June, 1847, I
was requested to go to Wilmslow, in Cheshire, to perform the
Cæsarean section in the case of a poor woman named Mrs.
Toft, who had been in labour since early on the morning of
the 12th—that was three days and a half—and for whom the
surgeons in attendance considered there was no hope of relief
otherwise than by such operation. I consequently started by
the first railway train, accompanied by my son, and arrived at
the house of the patient by half-past two o'clock.

" The patient was said to be about thirty years of age, and

had been married twenty-one months. She became pregnant shortly after marriage, but aborted at the third month. On the present occasion she had arrived at the full term of utero-gestation before calling for professional aid, which, indeed, she did only after labour had commenced, which was early on the morning of the 12th. Mr. Mayson, surgeon, of Wilmslow, had attended her alone, from Saturday morning till Monday morning, when he had Mr. Dean, surgeon, of the same place, associated with him, who continued his attendance along with Mr. Mayson up to the period when Mr. Dean came to request my aid.

"The patient had always been of a feeble constitution, with fair complexion ; but now she was excessively pale and ex-hausted, and was much disfigured by a large bronchocele. Her pulse was very rapid and feeble. On examination, the first object which attracted attention was the arm of a well developed child protruding from the vagina, proving it to be a case of shoulder presentation. The bones of the outlet of the pelvis were so crushed together that there was scarcely room for one finger to pass by the side of the protruding arm, so as to make an examination. The arm being pushed up, I ascer-tained that the rami of the pubes were so closely approximated that a finger placed edgeways could not reach the symphysis pubis ; and the tubera ischii were only about an inch apart, for there was no point where two fingers, placed side by side, could pass when placed transversely ; indeed, owing to the close approximation of the rami pubium, tubera ischii, and os coccygis, there was barely room sufficient to permit two fingers to pass the outlet of the pelvis in the antero-posterior direction. Owing to the shallowness of the pelvis, however, which must have been originally of small dimensions, I was the more readily enabled to reach and determine the dimensions of the brim. I ascertained that there was not as much available space in the antero-posterior direction as to permit the points of two fingers laid side by side to pass the brim of the pelvis,* except-ing about half an inch, exactly opposite the symphysis pubis, and there the fingers had barely room to pass. Beyond this, on either side, there appeared to be very little more than an inch of available space in the antero-posterior direction. The

* See Fig. 6th.

transverse diameter might exceed three inches; but then it was a crescentic form, which, of course, made it completely unavailable for delivery.

" My son having also made an examination of the patient, and made a similar estimate of the relative position of the bones of the pelvis at the brim and at the outlet, which were, moreover, firm and unyielding, we had no difficulty, in consultation with the other two surgeons, Mr. Mayson and Mr. Dean, in arriving at the conclusion that the woman must die undelivered if we did not instantly resort to the Cæsarean section. With such deformity as existed here in the pelvic bones, it must have been all but impossible to have broken down and extracted the fully developed child (which from the protruding arm this evidently was), even in a vigorous patient; but in a feeble woman like the one in question, exhausted to the last degree by the length of time she had been in labour, and the violence and acute suffering which she experienced from the pains, even up to the period when we were with her, it would have been perfectly futile and absurd to have made any attempt of the sort.

" The extreme violence and excruciating agony which the patient was suffering from the pains, rendered it the more desirable that the operation should not be unnecessarily delayed; and we therefore stated our views of the whole bearing of the case fully and fairly to the friends and obtained their consent, and subsequently the patient's also, when I proceeded to perform the operation in the usual manner at three o'clock. I deem it quite unnecessary to occupy your time by giving any details of the operation; for, although a formidable and a most important one, and one which ought only to be performed from the necessity of the case, still, *quasi* a surgical operation, it involves comparatively little difficulty to those well acquainted with the anatomy of the parts, and are in the frequent habit of operating. A very few minutes sufficed to make the necessary incisions and to extract the child, the placenta, and the coagula found in the uterus. All this —the stitching, dressing, and replacing the patient in bed— did not exceed ten minutes; and the whole pain sustained by the patient in consequence of the operation did not appear much to exceed a single pain, such as she had in our presence from the throes of Nature before we proposed the operation to her.

"The infant was large and well developed, but was dead, obviously from previous detachment of the placenta, for it was found quite detached, and surrounded by coagula, which at once accounted for the exsanguine appearance of the mother, as well as for the death of the child. There had been very little coloured discharge *per vaginam*, the egress having been completely closed by the shoulder of the child being impacted into the brim of the pelvis. Very little blood was lost by the incisions made during the operation, and very little passed *per vaginam* subsequently. After the operation, the patient seemed to suffer no more pain, but she passed quietly away, from exhaustion, five hours and a half after the operation.

"On the 17th I went over, accompanied by my son and another medical friend, for the purpose of having a post-mortem examination. Mr. Dean was also with us. We had almost been too late, as the company had assembled before our arrival for the purpose of interring the body. They consented to postpone it a very short time to allow us to make an inspection, but we were necessarily compelled to be very circumspect, as we were closely watched, and were thus prevented the opportunity of possessing ourselves of the pelvis. However, I had used the precaution of taking some plaster of Paris with us, and thus we were enabled to take an accurate model of the inlet of the pelvis. From this it was satisfactory to find that the estimate made of the brim of the pelvis, previous to undertaking the operation, had been very correct, as the following measurements of the cast prove :—About an inch immediately opposite the symphysis pubis, the diameter of the brim of the pelvis, from the symphysis pubis to the lumbar vertebræ, was one inch and three-eighths, and beyond that, on the left side, it abruptly diminished to nine-eighths of an inch, and from that to half an inch and nothing ; and on the right side to an inch and a quarter to an inch, and from that to the segment of a small circle. The transverse diameter, from right to left, was three inches and a half, but of this there was not more than two inches which would admit a ball to pass where it exceeded an inch in diameter. But, as already stated, the brim had assumed a crescentic form, so that when two straight parallel lines were drawn across the pelvis they only showed two inches by one inch as the largest available space. Under these circumstances, therefore, it is quite obvious that delivery, by mutilation of the infant, could not have been undertaken with any hope of success at any stage of the labour, however vigorous the patient might have been ; how much less so, then, with such a constitution as we had here to encounter, and with a shoulder presentation, too : the cavity of the pelvis was so small as not to admit a body greater than a lemon. At the left side and anterior part of the fundus uteri the walls of the uterus were fully an inch and a half in thickness, whilst a considerable portion on the posterior and right side was attenuated in an extreme degree ; so that from this circumstance, and the violent and cutting character of the pains witnessed by us, had she been left a short time longer undelivered, in all probability rupture of the uterus and death would have been the result. Very little blood had escaped *per vaginam* subsequent to the operation, and there was only a small clot found within the cavity of the uterus.

" Here, then, is a case which I think fully justifies the Cæsarean section ; for no medical man had been consulted until the woman was in labour at the full term of uterogestation, and with physical attributes which rendered it impossible for her to be delivered in any other manner than by the Cæsarean section. The only cause of regret is, that this had not been undertaken immediately after labour commenced ; for in that case there is every reason to believe that the life of the infant would have been spared, with a tolerable chance of safety for the mother also.

" This alternative had been proposed to the patient before I was sent for, but she obstinately held out against submitting to any such operation. I consider it but an act of justice to Mr. Dean and Mr. Mayson to record this fact ; and it took some management on my own part to obtain her consent at last. Although the operation failed to save the life of mother or child, still it relieved her of suffering for the last five hours and a half of her life, and her friends the pain of hearing her piercing screams, and witnessing the agonizing throes which accompanied the unavailing efforts of Nature to relieve her from her perilous condition.

"You can please append any remarks to this case which your experience in such cases may suggest to you.

"Believe me, my dear sir, yours very faithfully,

"JAMES BRAID.

"Arlington House,   "Dr. Radford, &c. &c.
    "May, 8th 1851.   "Manchester."

Dr. Radford's Remarks.—Malacosteon doubtless was the disease which caused the distortion of the pelvis of this poor creature; there are no data whereby to judge when its ravages on the bones commenced. Some of the contingent circumstances which happened, clearly prove the truth of those statements I made a short time ago. (Vide *London Medical Gazette* for April 4, 1851, vol. xlvii. p. 583.)

It was her first labour, and we have no evidence to show that any symptoms existed, either before or during her pregnancy, to induce her to place herself under medical treatment. The obstetrician was completely ignorant of the physical and organic condition of the pelvis until after labour had begun. It is then indisputably true that no other operation but the Cæsarean could possibly or safely be performed for her delivery. The great mischief of procrastinating the operation is emphatically proved by the result. Several serious evils arising from protracted labour are noticed by Mr. Braid. The death of the infant was undoubtedly produced by it, and most likely that of the mother. The internal flooding, the complete separation of the placenta, and the attenuated state of one portion of the structure of the uterus, which no doubt would have ended in its rupture, are solely to be attributed to delay.

In Cases 30 and 41 (Tab.), I have mentioned that violent irregular uterine contraction happened during the extraction of the infant, which, I stated, depended on the partial or complete detachment of the placenta; in the foregoing case, however, nothing of the kind took place, although this organ was lying loose in the uterus. Did the internal bleeding (which was so great as to bleach the general surface) act on the uterine tissue, and so influence its contractility? Was the absence of this spasm (when its supposed cause was present) owing to the extreme degree of attenuation of the uterine tissue?

CASE (Tab.) 53 (*Successful*).—The subject of this case—

Mary, wife of William Haigh—resides at Flats Fold, a mile from Ashton-under-Lyne, and about eight miles from Manchester. On my arrival, at half-past three o'clock p.m., I called upon Mr. Cluley, who accompanied me to the case, and, with the greatest courtesy and candour, gave me, as we passed along, the following particulars:—She had felt slight pains, according to the account of the friends, about a week; but Mr. Cluley thought that true parturient pains had only existed about three days, and which were so slight as not to require his interference. On this day (Sunday, May 20, 1849), at nine o'clock, he was again called, and although the pains were still trifling, he made an examination *per vaginam*, but was unable to feel either the os uteri or the presentation; he therefore had her taken out of bed and placed on the lap of a female friend, and again repeated his inquiry. The head of the infant was now felt, and the os uteri found dilated to the size of a half-crown piece. In this manœuvre he unintentionally ruptured the membranes. The pelvis, he mentioned, was considerably contracted.

I found her lying on the right side. Pulse 120; tongue clean and moist; her countenance tranquil, but a little flushed. Her bowels had been freely and fully moved this morning, and she had also freely and duly urinated; she was helplessly fixed on her side, and, when requested to turn, she remarked she suffered very great pain when she made an attempt to do so, or was by another person turned on the back. The pelvis was very considerably altered from its natural shape; its sides were flatter, and the posterior division of the ilia, especially on the left side, projected backwards, and the upper portion of the sacrum and the lower lumbar vertebræ had sunk in an inward and downward direction, so that a great concavity was perceived here. The uterus inclined rather to the right side, and stood considerably more forward than usual, although it had not resumed the retort form to the same degree as I have witnessed in former cases; its tissue felt soft and compressible. The fundus or upper division of the organ was fluctuant, and rounder in shape than it generally is after the discharge of the liquor amnii, which led me to conclude that a great portion of this fluid still remained. This opinion was corroborated when I attempted to ascertain the position of the infant; for at the

N

lower or cervical portion of the uterus, from whence it was
assumed the fluid had escaped, the projections of its body could
only be felt.

By a vaginal examination I found the lower aperture of
the pelvis very considerably diminished by the close approximation of the rami of the ischia and pubes, which nearly destroyed
the arch, and by their jutting forward there remained only a
narrow slit, which would not admit the point of a finger.   In
the transverse diameter two fingers could only be just placed
between the tubera ischii; the antero-posterior diameter was
also much shortened by the coccyx and the lower part of the
sacrum being considerably incurvated.   This great diminution
in the outlet rendered it difficult to measure the brim, so that
it was necessary to carry the hand very far backwards to
accomplish it.   Its figure was tripartite, or composed of three
divisions.   This alteration in the brim was occasioned by the
falling downwards and forwards of the upper part of the sacrum
and the lower lumbar vertebræ, which inclined a little more to
the left side, and by the body of the ossa pubis and ischii being
forced backwards and inwards, and by the jutting forwards of
the symphysis and rami of the pubes.*   The measurement of
the widest part of the conjugate diameter, in the two lateral
divisions, did not exceed an inch and a half; I could only place
two fingers, one lying a little over the other.   The anterior
division was not more than half an inch in its widest part, as
it would scarcely admit one finger edgeways.   The length of
this narrow opening is not relatively available in practice.   In
the transverse diameter of the brim I could just place three
fingers parallel with each other.   The external genitals were
free from tumefaction, and the vaginal lining was moist, and
of a natural temperature.   Whilst lying on her side, I was
unable to feel either the os uteri or the presenting part of the
infant; but on placing her on her back (which occasioned her
great pain), the os was felt to be dilated to rather more than
the size of a shilling.   She had not felt the movement of the
infant since the morning, but, by the stethoscope I satisfactorily heard the pulsation of its heart, which fact Mr. Cluley
afterwards corroborated.

With my opinion as to the position the Cæsarean section

* See Fig. 7th.

ought to take in obstetrics, I at once concluded that it was
the only operation that was justifiable, and indeed capable of
giving the best chance of life to both mother and infant.
Mr. Cluley most cordially acquiesced in this opinion. We
now informed her husband of the nature of the case, and the
means to be adopted. He answered, " if nothing else would
save her, he willingly submitted to any plan we considered
right." When a similar proposition was made to the poor
woman, she received it with the greatest resignation; it was
unaccompanied by either mental or physical disturbance. At
Mr. Cluley's request, Dr. Lees, Messrs. Hunt, Gibbons, Galt,
and Brewster were present at the operation. Before the
incision was made, I was anxious, as far as possible, to ascertain where the placenta was located, and I therefore placed my
ear over the left division of the uterus. From the negative
evidence, I concluded that it was not fixed on this side of that
organ. Mr. Cluley adopted the same plan, but thought he
heard the placental soufflet. I again applied my ear—still
heard nothing. Dr. Lees tried, and considered the sound to
arise from the friction of the ear on an interposed piece of
linen. Mr. Cluley, after a second trial, agreed in my opinion.
I therefore suggested the left side of the linea alba as the
situation to make the incision. I now raised the fundus uteri,
and Mr. Cluley divided the abdominal integuments on the left
side of the umbilicus to about six inches in extent, from which
very little blood was lost. An opening now was made into
the uterus by a scalpel, which was further extended upwards
and downwards by the probe-pointed bistoury. At this stage
some little bleeding took place from the divided sinuses, and
there was also a considerable discharge of liquor amnii. I now,
as quickly as possible, introduced one hand into the uterus,
over the infant's hip, and fixed the fingers under the flexed
thigh in the groin, and having placed the other hand on the opposite side of its breach, I extracted it vigorously alive. During
this manœuvre, the uterus strongly and regularly contracted.
The funis was now tied and divided by one of the gentlemen
present. Afterwards, I seized the funis with one hand, and
with the other readily detached and brought away the placenta,
which was fixed on the right latero-posterior surface of the
uterus. There was some blood discharged ; but not more than

frequently happens after ordinary or natural labour. Several convolutions of intestines, with a portion of omentum, now protruded, which had up to this time been supported and effectually restrained under the abdominal parietes, but they were readily returned. I carried my finger round the wound to ascertain if any portion of these viscera had descended into the uterus. The integuments were brought into proximity, and held together for a short time by a hand placed on each side; and, as there was no further discharge of blood, ligatures were inserted at an inch distance from each other. Mr. Cluley used a long needle with a scalpel-like handle for this purpose, which admirably answered: it is much superior to those in ordinary use. Straps of adhesive plaister were laid across the wound, and on each side a compress of lint was placed, and over all a bandage, just tight enough to give a firm support. During the whole time her mind was calm; she never even uttered a complaint. She remarked that her sufferings during the operation had been much less than what she had endured previously to it. Pulse from 80 to 90 in the minute. Tinct. opii ʒiss administered.

Half-past seven.—Pulse 100 to 120; dozing; there was no hæmorrhage or vomiting; had taken some gruel. Ordered mucilaginous beverages and farinaceous diet. At a later hour the same evening, Mr. Cluley saw her, and found a little abdominal uneasiness. She had slept and had a lochial discharge.

May 21st, Monday, half-past two p.m.—Pulse 130; tongue moist; face less flushed; abdomen tympanitic and slightly painful; fresh and plentiful lochia; bowels not moved; five ounces of water drawn by catheter; continue mucilaginous drinks, &c. An enema of warm water to be administered in the morning.

23rd, Tuesday, half-past nine a.m. Mr. Cluley had ordered forty drops of tinct. opii to be taken at bedtime. She had several times vomited a dark-coloured fluid during the night, and she still continues to do so; pulse 120; abdomen tympanitic, but not tender; tongue slightly furred; lochia natural; bowels still unmoved. After loosening the bandage, there was a discharge of sanious matter. Ordered an enema with ol. ricini ʒj; spt. terebinth. ʒij, &c. To take ext. col. co. gr. x; hyd. chlorid. gr. ij.

23rd, Wednesday, half-past nine A.M.—Hiccough has been troublesome; has bilious vomiting; tongue brownish; pulse 120; has a burning sensation in the throat, and the side of her mouth is excoriated; she says she tastes the turpentine which was given in the enema; bowels not moved. Ordered sodæ sub-borat. ʒij; aq. destillat. ʒiij; mist. acaciæ ʒiij. Capt. ʒj tertiis horis.—Gum-water to drink. To have an enema with three ounces of ox-gall and a pint of water.

24th, Thursday, half-past nine A.M.—Symptoms continue the same; but the tongue is slightly aphthous; two enemata were administered, which produced two small scybalous stools; wound much lessened in size, but its edges are flabby and have not united; ligatures still firm. To continue the same plan; to have the ox-gall enema repeated.

25th, Friday, half-past nine A.M.—All the symptoms better; has had free alvine evacuation. The integuments over the sacrum are inflamed and excoriated, and have a tendency to slough. To continue the same means. Ordered warm water enema. The parts over the sacrum to be dressed with collodion; warm water to be injected *per vaginam* into the uterus.

26th, Saturday, half-past nine A.M.—Tongue red and clean; bowels twice freely opened; has vomited several times since yesterday; wound granulating and looking well. To continue the same plan.

27th, Sunday, nine A.M.—Had suffered from occasional deafness and tinnitus aurium yesterday. This morning she is not so well. Pulse 125 and tremulous; the tinnitus aurium and deafness still continue; has numbness of one arm and leg; the bowels not moved. To take a little milk, to have a warm water enema administered, and afterwards one of milk. To take ammon. sesquicarb. if required.

28th, Monday, half-past nine A.M.—Yesterday, not so well; had delirious rambling. Pulse from 120 to 130; great restlessness and tossing about; was low in spirits; seemed much weaker; the bowels were moved; has taken the ammonia. She is much better this morning, and has had some sleep. The wound was patulous, and from it a dark-coloured and fetid fluid escaped. To continue milk diet; warm water to be thrown *per vaginam* into the uterus.

29th, Tuesday, half-past nine A.M.—All the symptoms are

better; the wound is filling up by granulation; one ligature came away. The bandage was wet, from the water which had been injected *per vaginam* into the uterus escaping through the wound. Slough over the sacrum came away, and the sores are looking well. Collodion to be again applied; to have a warm enema, water first, and afterwards one containing ox-gall, if required.

From the above date, up to June 7th, nothing occurred in the character of the symptoms to require particular comment. She continued progressively to improve. The wound gradually filled up by granulation, and is at this time nearly healed. The fistulous opening, through which the water which had been injected *per vaginam* into the uterus had escaped, is now completely obliterated. The sores over the back part of the sacrum, and on the nates, are also quite well.

During this period the diet chiefly consisted of milk; but towards the end of it animal food was allowed once a day. The mucilaginous mixture, with sodæ sub-boras, was the only medicine which she took, except the gum-water. When the bowels required relief, an enema of warm water was first administered, and, if necessary, this was followed by one containing ox-gall.

The collodion was continued as a dressing to the raw surfaces behind, until the latter part of the time, when pads of cotton, with a mild unguent, were substituted. During my absence from Manchester I received favourable reports of the patient from Mr. Gibbon, under whose professional care she was placed, in consequence of Mr. Cluley's severe illness. In his last letter, dated June 18th, he says: " The wound is very healthy, but not quite healed."

June 26th, Tuesday.—I visited her along with Mr. Cluley, and found her downstairs, and looking very well ; she remarked she was in excellent health. On removing the dressings, we found two or three spots of exuberant granulations, which only required the application of argenti nitras and a little dry lint.

July 15th, Sunday.—I called upon Mrs. Haigh. She was looking extremely well, and in excellent spirits. She observed she was better ; and could walk with more ease to herself than she could have done for a long time before the operation ; the wound was quite healed.

It was a great object with us that the infant's life should

be preserved. We therefore strenuously recommended a wet-nurse; and, if one could not be obtained, then that it should be supplied with asses' milk; but, from unavoidable circumstances, neither were procured until its life was placed in great danger. All those mischiefs consequent upon dry-nursing appeared: such as bowel affections, a threating of marasmus, and convulsions. At length a nurse was obtained, after which the infant improved, and on this day is quite well.

Before I proceed further, I take this opportunity of mentioning, the surgical part of the operation was most skilfully and dexterously performed by Mr. Cluley; and his punctual, assiduous, and unremitting attention to the patient are highly honourable to him. To him I am personally indebted, and return him my sincere thanks for his uniform great kindness and courtesy.

Remarks.—Mary Haigh was occupied before her marriage as a domestic servant, and was then strong, and capable of undergoing great exertion. She is of a sanguineo-lymphatic temperament; her skin fair, with a red blush on the cheeks; her hair of an auburn or reddish-brown colour; the tint of her eyes is rather peculiar, being of a brownish-grey, and they have an animated expression. Her father is now living and very healthy. Her mother has been dead many years; and, most likely, her death was occasioned by some chronic disease of the vertebræ, as I understood she was afflicted with abscess in the back.

Our patient is thirty-one years of age, and has been married nearly nine years. During this period she has had five children. The labours of the first four were natural and quick; the last of this number happened three years ago, and was so rapid that the infant was born before the obstetrician arrived. After the birth of the second, she was rather more delicate, and suffered a little from indigestion; and about five or six years since first complained of slight rheumatic pains about her hips. Two years since she was confined to bed for a short time by pains about the pelvis; but she gradually recovered, and afterwards was able to walk about tolerably well. Her general health remained the same up to the period of her last pregnancy. She was now observed to limp a little when she walked, and to be less in height.

During her gestation her progression was more difficult, and her gait more waddling. She also complained more of pelvic pains, and the diminution of her stature now evidently increased. Mollities ossium, the disease under which she suffered, usually commences during pregnancy, and generally becomes suspended in the interval, returning in an aggravated form in each successive pregnancy, until its ravages have completely destroyed the form of the pelvis. In this case, however, it did not exactly pursue this course. There is no doubt there existed a strong predisposition to the disease—most likely hereditary; and, probably, the disease began at the latter part of the second pregnancy; but, evidently, no great, if any, mischief was done to the pelvis at this time, or for a long time after this period, as the third and fourth labours were so rapidly and easily terminated. The rapidity of its progress is remarkable; for there is little doubt that the great degree of distortion took place immediately before and during the last pregnancy. Sometimes in this disease the bones are so soft that they yield when the hand is introduced to make an examination. This happened here, as Mr. Cluley thought he felt a giving way of the bones when he examined the pelvis.

Opium is generally given after great operations to lessen the shock on the nervous system; but, in the present instance, we had no evidence that such an effect existed, and therefore, on this account, the drug might have been omitted. A second dose was administered by Mr. Cluley, to which he attributed the vomiting which afterwards occurred. He considered that it had produced an effect similar to that which follows a debauch. It most likely constipated the bowels; but there is no doubt that this was chiefly caused by the bowel being compressed between the bulky uterus and the projection of the lower lumbar vertebræ and promontory of the sacrum. The garrulous delirium, the convulsive twitchings, and tinnitus aurium, &c., were considered by Mr. Cluley to depend on a state requiring more support: we, therefore, agreed to give a milk diet; and, as its effects were so satisfactory, it was continued to the end.

The negative system of treatment here pursued considerably contributed to the well-doing both of this case and also of the one in which I was concerned along with Mr. Goodman. I have also observed the same plan most beneficially carried out in the after-treatment of abdominal sections for the extirpation of large ovarian tumours. There are great objections to the use of purgatives after these great operations, as the mucous membrane of the bowels is so readily disturbed: we, therefore, only ordered two doses of pills, and trusted chiefly to the use of the enemata. The ox-gall enema was decidedly beneficial.

Rupture of the membranes, and evacuation of the liquor amnii, a long time before the operation, is always to be deplored; but, although this accident happened here, yet the great bulk of this fluid was still retained in the middle and upper portion of the uterus, which felt fluctuant and round in shape, and which admirably prevented the contraction of this organ, and so thereby lessened the chance of mischievous pressure on the maternal structures, and also contributed to the safety of the infant, and rendered its extraction more easy. The length of the uterine wound was also thereby diminished, in a degree proportional to the difference in the measurement of the uterine tissue, when distended by the contained fluid, and, after its evacuation, when shortened by contraction.

In the present case, happily, the water was not evacuated until a short time before the operation, and then only very partially; the pains were also fortunately so slight that no injurious pressure was made.

Sequel.—The following particulars of this highly valuable case will be especially interesting to most of my obstetric readers:—

Mary Haigh lived rather more than four years after she had recovered from the performance of the Cæsarean section, during which time she grew less in stature, and had great pain and difficulty in turning herself in bed. She was carried up and down stairs, and, as she was unable to move about, she was constantly confined to a sitting position. She was quite unequal to dress and undress herself. She suffered from pains about the pelvis. She, nevertheless, took great interest in, and partially attended to, her domestic duties until about twelve months before her death.

Her appetite continued good until a few months before her death, during which time she ate very little and had great loathing of her food, and had nausea. Her bowels were gene-

rally constipated, and she experienced great difficulty in void-
ing her stools. She had occasionally a slight cough, as from
cold. Menstruation continued regular nearly to the end of
life. She gradually lost flesh and declined in power. She
died on the 2nd of June, 1853.

Post-mortem Examination.—Her body was greatly emaciated.
The point of the sternum approximated very nearly to the
pubes, which jutted upwards; the spine was very short, and
incurvated outwards. The lower limbs were very thin, but
they were not shortened in length. On opening the abdomen,
the parietes were found very thin and attenuated. The viscera
were generally pale, but every organ appeared healthy. The
bladder was nearly empty, and was situated over the brim of
the pelvis. The uterus was of the usual size, and rested over
the brim of the pelvis. There was only a single band of lymph,
not thicker than a thread, passing from the anterior surface of
this organ to the peritoneum. The uterine tissue was uniform
in appearance, and there was not the slightest evidence to show
the site of the incision.

Pelvis.—The superior or false pelvis was considerably altered
in shape and lessened, and, indeed, it is nearly destroyed by
the descent of the lower lumbar vertebræ toward the pubes.
The expanded alæ ilii are so crushed together as to reduce the
concave venters to deep narrow sulci. The brim is trilobed in
shape, there being a very slight slit on each side of the sunken
lumbar vertebræ, and one on the anterior slit, which lies
between the rami of the pubes which juts out and forwards,
and are nearly approximated.*

Case (Tab.) 58 (*Unsuccessful*).—The following case I am
enabled to lay before my readers by permission of my esteemed
friend, Dr. Broughton, of Preston. He most liberally consented
for me to publish it, and has furnished me with his copious
notes. I shall, as literally as possible, give the case in
Dr. Broughton's own words :—

Ann, wife of Thomas Kenyon, of Higginson Street, Preston,
came to me as a charity patient on the 22nd of May, 1851.
Her sole occupation was her domestic duties. She was thirty-
one years of age, of a light complexion, and her eyes were blue.

* See Sketch, Fig. 8th.

She had suffered from a cough and felt very weak, and then she complained of pain in the back and about the hips, and a frequent inability to pass her water, except when lying on her belly across the foot of the bed. She waddled very much in her gait, and used a stick, and stated she had gradually been getting worse during the last two years. She said she was afraid she should not be able to walk about long unless she could obtain some relief. After a very strict inquiry, I could not ascertain that there was any reason to suppose there was the slightest hereditary predisposition to the disease (mollities ossium) she was now suffering from. I examined her very carefully, and found the mischief was chiefly confined to the pelvis.

At this time she felt uncertain in her mind whether she was pregnant; but she had some suspicion she was in that state, although her menstrual periods had been very irregular. I ascertained she had been pregnant six times before. Five of the labours were quite natural; in one, a few months previous to the present time, she aborted at the fourth month, and had afterwards suffered a great deal.

I made a vaginal examination, and very carefully endeavoured to find the uterus, but could not do so. I saw her several times, and desired her to let me know when she was quite certain about her pregnancy. On Saturday, July 5th, she called upon me, and stated she believed she was pregnant and had quickened the day previous. I again made a vaginal examination, but I was unable to find the os uteri.

I visited her on several occasions, but without obtaining any further information; and, therefore, I took my friend Mr. Haldan along with me to see her. He examined her, but could not find the os uteri.

The highly distorted state of the pelvis precluding any chance of attempting to induce premature labour by any artificial method, we unhesitatingly decided on the propriety of administering the ergot of rye. Full doses were at first given three times a day; but afterwards, each dose was more frequently repeated; and at last it was taken every two hours. Violent sickness occurred, but there was no uterine action excited, although this practice was continued rather more than a fortnight. At the expiration of this period, we felt afraid the fœtus could not pass through the pelvis, and therefore we decided not to interfere further, but concluded to attend to her general health. For this purpose we gave her cod-liver oil three times a day. She improved in health and strength, but had the same waddling gait. Her stature, according to her husband's account, was now upwards of four inches less than it was four years ago. She was visited from time to time, and her spirits were found uniformly good, although she was fully aware her child could not be born naturally. She had been directed to send on the very first indication of labour, and, on Wednesday, November 12th, at six P.M., her husband came to inform me. I sent him to request Mr. Haldan to meet me at the house immediately; we found she had very slight pains; and she stated she thought the liq. amnii had escaped; on a vaginal examination, no part either of the child or os uteri could be reached. The soft parts were moist and relaxed. Drs. Radford and Whitehead, of Manchester, having previously been aware of the case, had both most kindly said they would come and assist us; and, well knowing the value and importance of their opinions in such a case, I at once started for Manchester, during which time Mr. Haldan kindly consented to attend to the case.

They both accompanied me, leaving Manchester at half-past two A.M. On our arrival at Preston we at once proceeded to the house of our patient, and found she had not had very many pains. She was calm and tranquil. Her pains now recurred in regular succession. Drs. Radford and Whitehead made a very careful vaginal examination, and found the whole pelvis so extremely distorted, that they could not, after their most strenuous efforts, feel either the os uteri or ascertain the presentation of the child. They were both decidedly of opinion that it was quite impossible to deliver her by any other means than by the Cæsarean section.

I now sent for my friend Mr. Noble, who, after a careful examination, quite concurred as to the necessity of the Cæsarean section, as being the only available means for her delivery.

By the stethoscope we found the child was alive, but the pulsations of its heart were feeble and frequent, being fully 200 in the minute. The placental soufflet was heard over the anterior part of the uterus, and over a space extending to the fundus. No time was lost in commencing the operation. The abdominal parietes were divided by being first pinched up and then pierced by a straight scalpel, and the wound was afterwards enlarged by a button-pointed bistoury on the fingers. The uterus was now brought into view and was carefully opened. Its tissue was very thin, being very little thicker than strong paper (I would here bear my testimony to the statement of the late Mr. Barlow, of Blackburn, and which has been so much discussed). The incision was made directly over the placenta, as we had prognosticated ; it was quickly enlarged upwards and downwards by the bistoury on a director, and Dr. Radford most dexterously removed the child and placenta. The uterus rapidly contracted as it were *per saltum.* A large quantity of intestine protruded, but, by the valuable aid of Dr. Whitehead and Mr. Haldan, they were speedily returned and retained. She lost very little blood. The edges of the abdominal wound were now brought together by means of six sutures and long strips of emp. resina, which were supported by a large binder. Tinct. opii ʒj was given to her. She expressed herself as comfortable, and inquired about her child, which now cried lustily. She was under the influence of chloroform during the greatest part of the time ; we left her at seven A.M.

Ten A.M.—I found her restless ; her countenance was anxious, and there was a considerable oozing from the bottom of the wound ; her pulse is 132.

Twelve A.M.—She is still anxious. I found a large warm cloth on the abdomen, which I removed at once ; her pulse was 126. Tinct. opii ʒss was administered.

Four P.M.—She has vomited three times ; she had made water ; there was a slight lochial discharge. Her countenance is less anxious, and there is very little oozing from the wound ; her pulse is 124.

Six P.M.—She is better.

Nine P.M.—She is still better, and free from pain ; she has again passed a little water ; there is no lochial discharge since four P.M. Ordered pulv. opii gr. ij, to be taken immediately, and to be repeated at four A.M. if necessary.

14th, nine A.M.—She has vomited three times during the

night; she has had little sleep, and has made water freely; her pulse is 116.

Eleven A.M.—She is easy; her pulse is 118.

Half-past one.—She has vomited once; she is free from pain; her pulse is 124. Ordered pulv. opii gr. ij, to be immediately taken.

Four P.M.—She was made clean; her soiled bandages, &c., were removed and replaced by others; she says she has suffered less than after an ordinary labour. Her respiration is natural; pulse 124.

Half-past nine P.M.—Her pulse is 128; the respiration rather quick; her cheeks are rather flushed; she has again vomited. Liq. opii sed. gutt. xxv. She took a very little toast, and three tablespoonfuls of almost cold tea.

15th, ten A.M.—She has rested well; her pulse is 124; she has made water three times; the lochia are free. I now allowed her some weak beef-tea, made at my own house.

One P.M.—She is better, easy, and tranquil.

Four P.M.—She has had the beef-tea ordered this morning, and has slept a little; her pulse is 124; her countenance is placid, and her respiration natural.

Ten P.M.—Her husband has called, and says she is much worse. Mr. Haldan was then with me, and as I was very unwell, he very kindly saw her alone for me. He found her most anxious and restless; her pulse was 132; her breathing hurried; vomiting acid and bilious fluid. Mr. Haldan, with the long rectum tube, threw up some warm water, in which some soap was dissolved, but which was at once returned without any fæces. He ordered small doses of potass. bicarb., and a very small quantity of acid. citric., just enough to make it effervesce. She had an excellent night afterwards, and was quite easy at six A.M.

16th, ten A.M.—She is quite easy, and had not vomited. Pergat.

Three P.M.—On removing the dressings the edges of the wound were seen to gape between the sutures, and at the lowest part the intestines could be perceived. We brought the edges close together with the emp. ladani and emp. resinæ. The sutures were safe and firm. A clean binder and clean clothes were put on, and she was left quite comfortable.

Ten P.M.—She has had two loose stools, but is now easy; pulse 128; she says she is better than she had been previously.

17th, ten A.M.—She has been a good deal purged during the night, but is now quite free from pain; her stools are natural; the lochia free; she is less anxious. Ordered mist. cretæ co., with tinct. opii ♏xl, tinct. catechu f3ij.

Half-past three.—She is now much better, and free from pain; her bowels have only been very slightly moved since ten o'clock; her pulse is 116; she cannot take milk, which makes her sick; she wants a little barley gruel, which is allowed her; she says " has had a real after-pain," and is particularly cheerful; lochia abundant.

Five P.M.—She says she is very comfortable; pulse 116, and tongue quite clean; she remarked " I think I shall now get through I think." " I have suffered very little indeed."

Ten P.M.—About this time Mr. Haldan was sent for, as I had been very unwell for the last two days. He arrived at eleven, and found her quite altered in appearance; her pulse small and weak; her face pinched and pale, and, in fact, she had altogether changed. He was informed she had had a most violent attack of vomiting about eight o'clock, immediately after which she complained of some pain in the belly. On removing the binder, he found the wound open and some parts of the intestines protruding. He sent for me immediately, and on my arrival, about twelve o'clock, we made an attempt to return them, but we could not succeed, as more efficient aid was required than the nurse could afford. We then sent for Mr. Noble, who soon arrived, and we did all we could to replace the protruded bowels and close the wound. We, however, could not succeed, and as it was quite evident she was sinking, at her request we desisted, and covered them again with a warm moist cloth, and left her, receiving her dying thanks for our attention. She sank at a quarter to six A.M.

At twelve o'clock we went to examine the body and replace the intestines. We found them as we had left them (at one o'clock A.M.), protruding, but much altered in appearance. These viscera were now glued by plastic lymph, and a part of the ilium appeared to be gangrenous, and the omentum adherent to the small intestines. Those bowels which remained

in the abdomen looked healthy, with the exception of the
ascending colon, which adhered to the fundus of the uterus;
on cutting into the uterus it was found healthy and natural.
The wound in it was partially closed.   The liver was pale,
large, and encroached on the chest.   There were old bands of
adhesion between this organ and the diaphragm, and also
between it and the walls of the abdomen.   The stomach was
distended with flatus, but it was healthy.   The heart was large,
flabby, and pale.   The right lung collapsed and adherent to
the pleura costalis by old bands of adhesion.   Its lower part
was hepatized and red.   The left lung was free, small, and
full of frothy mucus.   Its lower part was congested at its
apex ; there was grey deposit ; the bladder was healthy.   The
head was not examined.   The brim of the pelvis,* on the
right side, was so contracted from the falling forwards of the
sacrum, as barely to admit one finger, and anteriorly at the
pubes rather less.   On the left side there was a little more
room ; in one part, near the sacro-iliac synchondrosis, it just
admitted two fingers in the antero-posterior diameter, but only
one in the lateral.   In fact it was a notch ; the sacrum
appeared to have come forwards from the ilium by the elongation of the ligaments.   The rectum passed here.   In no part
could a ball one inch and one-third pass.   The ascending rami
of the ischia were so closed as to barely admit one finger,
and the tuberosities are very nearly approximated.   The os
coccygis is thrown forwards.   Thus the cavity of the pelvis was
encroached upon to a very remarkable degree, rendering the
introduction of even a very small hand quite impossible, and the
longest finger could do little more than reach the promontory of
the sacrum.   The urethra was drawn upwards and backwards
behind the symphysis of the pubis, requiring a very long and
curved catheter.   The situation of the urethra accounts for
the difficulty she had in emptying the bladder.   I do not
doubt but the uterus was retained above the superior aperture
by the state of the bones.   Her husband informed me "she
had been growing together during the last two years, and that
she gradually grew worse and closer together.   The child is
now alive and quite well, and under a wet nurse."

    Dr. Radford's Remarks.—In the privilege given to me of

---

* See Fig. 9th.

making a few remarks on Dr. Broughton's case, I cannot allow the opportunity to pass without expressing my high appreciation of the great humanity and kind attention to the interest of his (a hospital) patient, and for the very ardent desire he showed, not only to bestow on her all the advantage within his reach in Preston, but in travelling sixty miles to obtain further aid. Throughout the entire case his treatment was very judicious, and when he had decided upon the propriety of allowing her to go on to the end of pregnancy, he carefully watched her, and urgently pressed upon her and her husband the necessity of acquainting him as early as possible after labour had commenced. My remarks need only be brief, as I have already, in the preceding observations, treated upon most of the interesting points which are contained in both this and some of the other cases.

This poor woman had been naturally delivered of five full-grown infants—there had been no artificial aid required—and therefore all the pelvic mischief must have been produced after the last labour. The disease had made rapid progress, having commenced about two years before. Its ravages were very great and extensive in the pelvis, involving the brim, cavity, and outlet.

It is quite unnecessary for me to go into the physical character of a pelvis distorted by mollities ossium (and the one belonging to Dr. Broughton's patient was extremely so), to show the difficulties the practitioner has to encounter to ascertain such information as ought to guide his judgment of the case. In this case, Dr. Broughton and his medical friends were all unable, when first consulted, and also afterwards, to find the os uteri ; and therefore he most wisely decided not to attempt to induce abortion by any artificial means ; if he had done so, he might have produced fatal mischief. The uterus and bladder, in such cases as this, are situated above the brim of the pelvis, and lean very much forwards over the beak-like projection of the approximated rami of the pubes and ischia. Whatever may be said, or even thought, as to the propriety of the induction of premature labour or of abortion in such cases as this, I most positively deny its possibility ; to make an attempt would be most unwarrantable. (See observations, *ante*).

This poor creature was greatly afflicted, and her vital powers at a low point, as nearly all those women are who suffer under such a constitution and a local disease.

With the exception of these conditions, her case was one favourable for the operation. Dr. Broughton had taken care that her labour was not too much protracted, and the character of the pains were favourable to the issue—being weak and irregular—so that the head of the infant was not forcibly pressed upon the maternal tissues. It is true the membranes had ruptured at the commencement of the labour, but there had been no mischief done. The employment of the stethoscope had warned us of the position of the placenta, but it was quite impossible to avoid this organ in making the incision, and it was therefore consequently cut upon. By this foreknowledge of the certainty of the placenta being in the way of the incision, I was prepared instantly to pass the right hand and seize one lower extremity of the infant, whilst I passed the left hand downwards so as to embrace its head, and so, by a compound movement, to throw out and raise from the uterus the body of the infant, which lay obliquely, its head being on the brim of the pelvis in the first position. By this rapid extraction I doubt not the infant escaped the grasp of the uterus. This result happened in two of my cases, and has been adverted to in the previous observations. The abdominal parietes were very thin and attenuated, and therefore every care was taken to have them well approximated and supported by ligatures, &c.; but, notwithstanding all the precautions taken, the parts yielded, and some portions of the bowels protruded, and could not be replaced. This most unfortunate event happened at the end of the fourth day after the operation, and doubtless was produced by the violent attack of vomiting which then took place. Before this occurrence, everything promised a favourable issue. She had most certainly vomited on the day of the operation, and again twice on the day after its performance. Chloroform was administered at the commencement of the operation, and she was kept under its influence "during the greatest part of the time" of its performance. Was the vomiting in any degree owing to the chloroform? (See preceding observations, p. 25.)

Dr. Broughton, in a note, December 15, 1851, says: "The

case to me has been a most instructive one, and in all future operations in which a large opening is made in the abdomen, I will use metallic ligatures, and replace at alternate spaces fresh ones at the first dressing ; I shall then secure the wound." This suggestion is highly deserving of attentive consideration, and in my opinion its adoption promises to be of great use. It is most important, if possible, to secure the edges of the wound, as there are recorded in the Tables other cases in which death resulted in consequence of their giving way. The case of Mrs. Sankey (although she recovered) is an excellent example of the disastrous risks which are produced by such an accident.

TABLE I.

OF RECORDED CASES OF CÆSAREAN SECTION

No.	Year.	Name and residence of the patient.	By whom and where the case is related.	Operator.
1	Jan. 9, 1738.	Alice O'Neil, aged 33 years, near Charlemont, Ireland.	Mr. Duncan Stewart, *Edinburgh Essays*, vol. v. p. 439.	Mary Donnally.
2	June, 1757.	Patterson, Canongate, Edinburgh.	Smellie's *Midwifery*, vol. iii. coll. 39, No. 2, p. 373.	Mr. Smith.
3		Not named.	Manuscript Lectures.	Professor Young.
4		Not named.	Manuscript Lecture.	Professor Young.
5		Not named.	Mentioned in Dr. Hamilton's *Outlines of Midwifery*.	Mr. Alex. Wood, Edinburgh.
6	Before 1740.	Not named, Rochdale, Lancashire.	Dr. Hull's *Defence*, p. 67.	Dr. White, Manchester.
7	Oct. 1769.	Martha Rhodes, London.	Dr. Cooper and Mr. Henry Thompson, *Lond. Med. Obs. and Inquiries*, vol. iv.	Mr. H. Thompson.
8	1774.	Elizabeth Clerk, aged 30, Edinburgh.	Dr. Alex. Hamilton, *Outlines of Midwifery*, p. 293.	Mr. W. Chalmers, Edinburgh.
9	August, 1774.	Elizabeth Forster, London.	Dr. Cooper, *Lond. Med. Obs. and Inq.*, vol. v.	Mr. Hunter, London.
10	1775.	Not named.	Dr. Hull's *Defence*, p. 66.	Mr. W. Whyte, Glasgow.
11	1777.	Elizabeth Hutchinson, aged 40, Leicester.	Dr. Vaughan, *Cases and Observations on Hydroph.*	Mr. Atkinson, Leicester.
12	Nov. 1793.	Jane Foster, aged 40, Blackrod, Lancashire.	Mr. Barlow, *Medical Records and Researches*, p. 154: also, his *Observations*, p. 355.	Mr. Barlow.
13	Sept. 1794.	Isabel Redman, aged 33, Blackburn.	Dr. Hull's *Defence*, p. 172.	Dr. Hull.
14	June, 1795.	Jean Douglas, Edinburgh.	Dr. Alex. Hamilton, *Outlines of Midwifery*, p. 299.	Dr. James Hamilton, jun.
15	Sept. 1793.	Anne Lee, Manchester.	Dr. Hull's *Defence*, p. 162.	Dr. Hull.
16	1798.	Janet Williamson, aged 38, Kirriemuir, Forfarshire.	Dr. Hull's *Defence*, p. 188.	Mr. Kay, Forfar.

TABLE I.

IN GREAT BRITAIN AND IRELAND.

Cause of difficulty.	Duration of the labour.	Mother.		Child.		Mother survived.
		Preserved.	Died.	Preserved.	Died.	
Not stated.	12 days.	P.			D.	
Distorted pelvis, most likely from mollities ossium.	7 days.		D.		D.	18 hours.
Distorted pelvis, from rickets.	No account.		D.	P.		
Distorted pelvis, from mollities ossium.	No account.		D.	P.		3 days.
Not stated.	No account.		D.		D.	
Not stated.	No account.		D.		D.	
Distorted pelvis, from rickets.	Nearly 30 hours.		D.		D.	5 hours.
Distorted pelvis, most likely from mollities ossium.	12 days.		D.	P.		26 hours.
Distorted pelvis, from mollities ossium.	60 hours.		D.	P.		25½ hours
No account.	No account.		D.		D.	No account.
Distorted pelvis, from mollities ossium.	Nearly 3 days.		D.	P.		About 80 hours.
Distorted pelvis, from fracture.	5 days.	P.			D.	
Distorted pelvis, from mollities ossium.	12 hours.		D.	P.		3½ hours.
Distorted pelvis, from mollities ossium.	5½ hours. Spurious pains three nights previous.		D.		D.	33 hours.
Distorted pelvis, from rickets.	10 days.		D.		D.	6 hours.
Distorted pelvis, from mollities ossium.	More than 3 days.		D.		D.	11 days.

TABLE I. (*continued.*)

No.	Year.	Name and residence of the patient.	By whom and where the case is related.	Operator.
17	June, 1799.	Elizabeth Thompson, aged 32, Hazelhurst, near Ashton-under-Lyne, Lancashire.	Mr. Wm. Wood, *London Med. Memoirs*, vol. v.	Mr. Wood, Manchester.
18	March, 1800.	Not named; resided at Edinburgh.	Sir Charles Bell, *Lond. Med.-Chir. Trans.*, vol. iv.	Mr. Jno. Bell.
19	August, 1801.	Hannah Rheubotham, aged 41, Manchester.	Mr. Wm. Wood, *Lond. Med. Phys. Journ.*, vol. vi. p. 346.	Mr. W. Wood, Manchester.
20	Feb. 1801.	Susan Holt, aged 36, Lower Shore, Rochdale, Lancashire.	Dr. Hull's *Translation of Baudelocque*, p. 134.	Mr. Walter Dunlop.
21	July, 1811.	Mrs. M., Leith, Scotland.	Dr. Kellie, *Ed. Med. and Surg. Journal*, vol. viii.	Dr. Kellie.
22		Wife of Benjamin Buckley, Staleybridge, Lancashire.	Not before mentioned.	Mr. Hutton.
23		Name unknown; resided at Staleybridge, Lancashire.	Not mentioned before.	Mr. Hutton.
24	August, 1814.	Wife of James Tinker, aged 34, Moston, Lancashire.	Mr. K. Wood, *Med.-Chir. Trans.*, vol. vii. p. 264.	Mr. K. Wood.
25	Jan. 1817.	Wife of Wm. Ratcliffe, aged 35, Staleybridge, Lancashire.	Not mentioned before.	Mr. Hutton.
26	July, 1817.	Ann Hacking, aged 42, Blackburn, Lancashire.	Mr. Barlow, *Observations*, p. 361.	Mr. Barlow.
27	April, 1820.	Mary Ashworth, aged 42, Denton, Lancashire.	Dr. Radford, *Edin. Med. and Surg. Jour.*; *Prov. Med. and Surg. Jour.*, vol. xv.; *Lond. Med. Gaz.*, vol. xlviii. p. 95.	Mr. Morris, Ashton.
28	Sept. 1820.	M:s. Lowe, aged 30, Perth, Scotland.	Dr. Henderson, *Ed. Med. and Surg. Jour.*, vol. xvii. p. 105.	Dr. Henderson.
29	April, 1821.	Wife of G. Ridgedale, aged 42, Blackburn, Lancashire.	Mr. Barlow, *Observations*, p. 375.	Mr. Barlow.
30	May, 1821.	Mary Nixon, aged 39, Manchester.	Dr. Radford, *Edin. Med. and Surg. Journ.*; *London Med. Gaz.*, vol. xlviii. p. 98; *Prov. Med. and Sur. Jl.*, vol. xv. p. 426, 1851.	Mr. Wilson.

TABLE I. (*continued.*)

Cause of difficulty.	Duration of the labour.	Mother.		Child.		Mother survived.
		Preserved.	Died.	Preserved.	Died.	
Distorted pelvis, from mollities ossium.	21 hours.		D.	P.		76 hours.
Distorted pelvis, from mollities ossium.	Not stated.		D.	P.		Very short time.
Distorted pelvis, from mollities ossium.	61 hours.		D.		D.*	24 hours.
Distorted pelvis, from mollities ossium.	56 hours.		D.	P.		6 days and 9 hours.
Distorted pelvis, from mollities ossium.	About 36 hours.		D.	P.		About 24 hours.
Distorted pelvis, from mollities ossium.	3 days.		D.		D.	
Distorted pelvis, from mollities ossium.	No account as to precise time; but long.		D.	P.		
Distorted pelvis, from mollities ossium.	About 40 hours.		D.		D.	10 hours.
Distorted pelvis, from mollities ossium.	No account.		D.	P.		48 hours.
Distorted pelvis, from mollities ossium.	13 hours.		D.	P.		76 hours.
Distorted pelvis, from mollities ossium.	37 hours.		D.		D.*	35 hours.
Distorted pelvis, from mollities ossium.	102 hours.		D.	P.		20 hours.
Distorted pelvis, from mollities ossium.	About 34 hours.		D.	P.		52 hours.
Distorted pelvis, from mollities ossium.	22 hours.		D.		D.	67½ hours.

* Dead before operation.

TABLE I. (*continued.*)

No.	Year.	Name and residence of the patient.	By whom and where the case is related.	Operator.
31	April, 1826.	M. R., aged 22, Stobsmuir, Scotland.	Mr. Crichton, Dundee, *Ed. Med. and Surg. Journ.*, vol. xxx. p. 53.	Mr. Crichton.
32	August, 1826.	Mary Forrest, aged 35, three miles from Blackburn, Lancashire.	Mr. Barlow, *Lond. Med. and Surg. Journ.* vol. iv.	Mr. Barlow.
33	Sept. 1829.	Mrs. M., aged 26, Belfast.	Communicated by Dr. W. Campbell, *Edin. Med. and Surg. Journ.* v. xxxv. p. 351.	Dr. M'Kibbin.
34	Nov. 1834.	Mrs. ——, Dublin.	Dr. Montgomery, *Dublin Med. Journ.*, vol. vi.	Mr. Porter.
35	April, 1834.	Mary Bamford, aged 38, Great Easton, Rockingham.	Mr. T. L. Greaves, *Lancet*, vol. ii. 1833-4.	Mr. Greaves.
36	May, 1835.	Sarah Bate, aged 36, Birmingham.	Mr. Knowles, *Trans. Prov. Med. Assoc.*, vol. iv. p. 376.	Mr. Knowles.
37	Aug. 14, 1840.	Mary Ann Jones, aged 39, Manchester.	Mr. James Whitehead, Manchester, *Med. Gaz*, vol. xxviii. p. 939, 1840-41.	Mr. Whitehead.
38		Name not given, or age; private note, aged 30.	Mr. Dendy, Medical Society of London, *Lancet*, 1842-43, vol. xliii. p. 691.	Mr. Bryant, Lambeth.
39	Oct. 17, 1842.	Mary Davis, aged 23, Reading.	Mr. T. B. Hooper, Reading Medical Society, *Lancet*, 1843, vol. xliii. p. 689.	Mr. Hooper.
40	March 8, 1842.	Mary Jepson, aged 43, a weaver, Darwen.	Mr. S. H. Wraith, Darwen, *Prov. Med. Jour.*, vol. v. p. 329, 1842-43.	Mr. Wraith.
41	Feb. 22, 1843.	Mary Forrest, aged 38, a weaver, Stockport.	Dr. Radford, *Med. Gazette*, vol. xlvii. p. 801; *Prov. Med. and Sur. Jl.*, vol. xv. p. 287, 1851.	Dr. Radford.
42	July, 1825.	Betty Wilcock, aged nearly 49.	Messrs. Hardy and Bailey, *Ass. Med. Jl.*, vol. iv. p. 45, 1856.	Mr. Bailey.
43	Oct. 18, 1837.	E. Hull, aged 25, Sunderland.	Mr. J. Ward, *Lond. Med. Gazette*, vol. xxi. p. 817.	Mr. J. Ward.

TABLE I. (*continued.*)

Cause of difficulty.	Duration of the labour.	Mother. Pre-served.	Mother. Died.	Child. Pre-served.	Child. Died.	Mother survived.
Distorted pelvis, from fracture, &c.	6 days.		D.	P.		8 hours.
Distorted pelvis, from mollities ossium.	From 30 to 36 hours		D.	P.	D.	More than 3 days.
Distorted pelvis, from a large exostosis arising from the sacrum.	About 30 hours.		D.		D.*	17 hours.
Fibrous tumour, growing from substance of uterus and covered by peritoneum.	18 to 20 hours.		D.		D.*	21½ hours.
Distorted pelvis, from mollities ossium.	About 34 hours.	P.		P.		
Distorted pelvis, from mollities ossium.	About 30 hours.	P.		P.		
Distorted pelvis, from mollities ossium.	24 hours; in active labour for 2 to 3 hours.	P.		P.		32 days 10 hours.
Distorted pelvis, by rickets.	24 hours.		D.	P.		60 hours.
A large tumour, arising from the sacrum.	3 days.		D.		D.*	40 hours
Distorted pelvis, from malacosteon.	10 hours.		D.		D.	3 hours.
Distorted pelvis, from malacosteon.	53 hours.		D.		D.	About 27 hours.
Distorted pelvis, from mollities ossium.	83 hours.		D.	P.†		61½ hours.
Distorted pelvis, from (presumed) rickets.	27 to 30 hours.		D.		D.*	5 days 7½ hours.

* Dead before operation.   † Two.

TABLE I. (*continued.*)

No.	Year.	Name and residence of the patient.	By whom and where the case is related.	Operator.
44	1840.	...	Dr. Churchill's *Operative Midwifery*, Table, p. 205. Communicated in a letter to Dr. C.	Dr. Elliott, Waterford, Ireland.
45	Feb. 1842.	Helen McKenzie, aged 35.	Mr. Alex. Ross, *London and Edin. Monthly Journal*, vol. ii. p. 425.	Mr. A. Ross, Invergordon.
46	August, 1844.	Rebecca Brooks, aged 27, Welford.	Mr. Fred. Cox, *Prov. Med. and Surg. Journal*, vol. viii. p. 382.	Mr. F. Cox, Welford.
47	Feb. 21, 1845.	Wife of Richard Instan, aged 40.	Mr. J. Milman Coley, Pamphlet.	Mr. Coley.
48	March, 1845.	Mrs. R., Shettlestone.	Mr. William Lyon, *Lon. and Edin. Jour. of Med. Science*, vol. v. p. 885.	Mr. Lyon.
49	Nov. 1845.	Mrs. Sankey, aged 41, Salford.	Dr. Radford, *Lond. Med. Gaz.*, vol. xlvii. p. 894; *Prov. Med. and Sur. Jl.*, vol. xv. p. 315, 1851.	Mr. Goodman, assisted by Dr. Radford.
50	Jan. 1847.	Sarah Bartlett, aged 37, removed for operation to St. Bartholomew's Hospital, London.	*Lancet*, 1847, vol. i. p. 139.	Mr. Skey, London.
51	June, 1847.	Mrs. Toft, aged 30.	Dr. Radford, *Lond. Med. Gazette*, vol. xlviii. p. 238; *Prov. Med. and Surg. Jl.*, vol. xv. p. 483, 1851.	Mr. Jas. Braid, Manchester.
52	1849.	Mrs. Rogers, aged 40, six miles from Lisburn, in a mountainous district.	Mr. John Campbell, Lisburn, Ireland, *Lond. Med. Gazette*, vol. xliii. p. 1105.	Mr. Campbell.
53	May, 1849.	Mary Haigh, aged 31, near Ashton-under-Lyne.	Dr. Radford. Read at Worcester, at Meeting of Prov. Med. Assoc., Aug. 1, 1849. *Prov. Med. & Surg. Jl.*, afterwards in *Lond. Med. Gaz.*, vol. xlvii. p. 1110.	Mr. Cluley, assisted by Dr. Radford.

TABLE I. (*continued.*)

Cause of difficulty.	Duration of the labour.	Mother.		Child.		Mother survived.
		Preserved.	Died.	Preserved.	Died.	
Distortion of pelvis; kind not mentioned.	...		D.		D.	
Distortion of pelvis, most probably mollities ossium (outlet contracted). Tuberosities of ischia approximated very closely; coccyx closing up lower part.	Midwife stated she had been in labour, more or less, for 12 days. Much time elapsed after Mr. Ross decided on necessity of operation.		D.	P.		5 days 7 hours.
Distortion of pelvis, most likely rickety, but not stated.	30 to 40 hours.		D.		D.*	54 hours.
Distortion of pelvis, said to be rickety.	At least 10 days.		D.		D.	9 days.
A large tumour, the size of a child's head.	72 hours.		D.	P.		36 hours.
Distortion of pelvis, from mollities ossium.	12 to 14 hours.	P.		P.		Lived.
Distortion, from rickets.	5 hours 5 minutes.		D.	P.		36 hours.
Distorted pelvis, from mollities ossium.	3 days, and 10 to 12 hours.		D.		D.†	5½ hours.
Distorted pelvis, from mollities ossium; clearly so, from the account of the case.	52 hours.		D.	P.		7 days.
Distorted pelvis, from mollities ossium.	Slight pains for two or three days. Membranes unruptured until a few hours before operation.	P.		P.		

* Craniotomy.   † Dead before operation.

TABLE I. (*continued.*)

No.	Year.	Name and residence of the patient.	By whom and where the case is related.	Operator.
54	May, 1850.	Elizabeth Williams, in the 27th year of her age.	Dr. Charles West, *Med.-Chir. Trans.*, vol. xxxiv. p. 61.	Mr. Skey.
55	May, 1850.	Mrs. Kennaway, aged 43.	Mr. M. Nimmo, Dundee, *Mon. Jour. Med. Science*, Sept. 1850, p. 226.	Mr. Nimmo.
56	Sept. 1850.	Sarah ——, Bethnal Green, aged 23.	Dr. Henry Oldham, London, *Med.-Chir. Trans.*, vol. xxxiv. p. 80.	Mr. Poland, London.
57	June, 1851.	Sarah L., aged 28.	Dr. Henry Oldham, *Guy's Hosp. Reports*, vol. vii. p. 426.	Mr. Poland.
58	Nov. 1851.	Ann Kenyou, aged 31.	Dr. Broughton, Manuscript.	Dr. Broughton, Preston, assisted by Dr. Radford and Dr. Whitehead.
59	1853.	Aged 41.	Dr. Charles Waller, *Medical Times and Gazette*, vol. vi. p. 266.	Mr. Le Gros Clark.
60	Feb. 28, 1848.	Mrs. Y., Shaftesbury, Dorset, aged 42.	Mr. R. W. Sanneman, Chelsea, *Lancet*, vol. ii. 1850, p. 50.	Mr. Sanneman.
61	June 24, 1854.	Mrs. ——, Cupar, Scotland.	Professor Simpson, Edinburgh, *Assoc. Med. Jour.*, vol. ii. 1854, p. 1066.	Professor Simpson.
62	1854.	Mary A. Johnson, aged 31.	Paper read at Royal Med. and Chir. Soc., April 13, 1858; from Notes kindly furnished by Dr. Greenhalgh to me.	Dr. Greenhalgh.
63	1854.	Lydia Lowdey.	Notes kindly furnished me by Dr. Greenhalgh.	Dr. Greenhalgh.
64	Dec. 5, 1854.	Martha C., Nottingham.	Dr. J. C. L. Marsh, *Lancet*, vol. ii. 1863, p. 560.	Dr. Marsh.

TABLE I. (*continued.*)

Cause of difficulty.	Duration of the labour.	Mother.		Child.		Mother survived.
		Preserved.	Died.	Preserved.	Died.	
Distorted pelvis, from mollities ossium.	16 hours.		D.	P.		108½ hours.
Distorted pelvis, from mollities ossium.	5 days. Occasional pains; but had regular labour for 17½ hours.		D.	P.		Scarcely 3 hours.
Distortion of the pelvis from rickets.	84 hours from the time labour was artificially induced.		D.		D.*	About 44 hours.
Cancer of the cervix uteri.	Indefinite slight uterine contraction for 2 or 3 days. Membranes ruptured about 12 hours before operation.	.P		P.		Recovered from.
Distorted pelvis, from mollities ossium.	11 hours.		D.	P.		5 days 2 hours.
Large fibrous tumour.	30 hours. Membranes ruptured 78 hours before operation.		D.	P.		36 hours.
Distortion of the pelvis, from mollities ossium	12 hours.		D.		D.	23 hours.
Distortion of the pelvis, from mollities ossium.	60 or 70 hours.		D.		D.†	
Distortion of the pelvis, from mollities ossium.	About 6 hours.		D.	P.		3 weeks.
Distortion of the pelvis, from rickets.	12 hours.		D.	P.		4 days.
Distortion of the pelvis, from mollities ossium.	55½ hours.		D.	P.		48 hours.

* Head perforated.　　　　　† Living; died soon after.

TABLE I. (*continued.*)

No.	Year.	Name and residence of the patient.	By whom and where the case is related.	Operator.
65	Feb. 29, 1856.	Mrs. Runham, Lawton, aged 30.	Mr. Humphry, *Assoc. Med. Jour.*, vol. iv. 1856, p. 779.	Mr. Humphry
66	Feb. 25, 1856.	Nancy Nixon, aged 27, a travelling hawker, in a miserably low cellar, Staleybridge, Lancashire.	Dr. Chas. Clay, *Midland Quarterly Journal of the Medical Sciences*, Part I. p. 21.	Dr. Clay.
67	Oct. 1, 1856.	Anne N.	Dr. W. H. Thornton, *Lancet*, vol. i. 1857, p. 313.	Dr. Thornton.
68	Feb. 19, 1858.	Matilda T., aged 20, Newport, Monmouthshire.	Mr. James Hawkins, *Lancet*, vol. i. 1858, p. 529.	Mr. Hawkins.
69	July 12, 1858.	Mrs. N., aged 30, Harrington, brought to University College Hospital.	Dr. Murphy, *Dub. Quar. Jour. of Med. Sc.*, vol. xxvii. (new series), p. 108.	Mr. Quain.
70	Dec. 1859.	Mrs. H., Walton-le-dale, near Preston, Lancashire.	Dr. H. Ashton, *Lancet*, vol. i. 1860, p. 440.	Dr. Ashton.
71	Dec. 10, 1860.	Emma P.	Dr. James Edmunds, *Lancet*, vol. i. 1861, p. 4.	Dr. Edmunds.
72	Feb. 2, 1861.	Isabella King, unmarried, aged 23, Aberdeen, Scotland.	Dr. Robert Dyce, *Edin. Med. Jour.*, vol. vii. p. 895.	Dr. Dyce.
73	Aug. 7, 1862.	E. M., unmarried. aged 17, Tottenhall.	Dr. David Johnson, *Lancet*, vol. ii. 1862, p. 475.	Dr. Johnson.
74	Dec. 24, 1862.	Mary Ann ——, unmarried, aged 42, Kingswood, Bristol.	Dr. J. G. Swayne, *Obstetric Trans.*, vol. v. 1863, p. 84.	Mr. Coe, at the Bristol General Hospital.
75	1863.	Eliza Hubbard, St. Bartholomew's Hospital.	Dr. Greenhalgh, private communication.	Mr. Skey.

* Embryotomy had begun ; the arm removed.   † Alive when extracted ; but died in a few seconds.
‡ Dead before operation.

TABLE I. (*continued.*)

Cause of difficulty.	Duration of the labour.	Mother.		Child.		Mother survived,
		Preserved.	Died.	Preserved.	Died.	
Distortion of the pelvis, from mollities ossium.	Not definitely stated; but most likely about 30 hours.		D.		D.*	20 hours.
A large tumour of a firm fibro-cartilaginous texture.	3 days.		D.		D.†	19 days.
A bony projection from promontory of sacrum, evidently exostosis; perhaps cellulated.	About 18 hours.	P.			D.‡	
Distortion of pelvis; kind of deformity not stated; but as lameness and inability to walk in early life, it was most probably from rickets.	Not definitely stated.	P.		P.		Lived.
Distortion of the pelvis, from mollities ossium.	As far as can be computed, 4 days.		D.		D.	Nearly 48 hours.
Distortion of the pelvis, from mollities ossium.	As far as can be ascertained from the data, about 17 or 18 hours.		D.	P.		25 hours.
Hard cancer of os and cervix uteri.	Fully 6 days.	P.		P.		Lived.
Distortion of the pelvis, from rickets.	As far as can be computed, 4 to 5 days.		D.		D.§	43 hours.
Small underdeveloped pelvis, perhaps rickety. Antero-posterior diameter only ascertained *per vaginam* to be 2½ in. Preternatural position of child, right hand and the two feet.	12 hours, when first seen by Dr. J.; afterwards the time in his attempt to deliver, &c., 17 hours.		D.		D.‖	46 hours.
Distortion of the pelvis, from congenital malformation. It was like that produced by rickets.	60 hours before she came to the hospital, and a few hours after her arrival.		D.	P.		42 hours.
Medullary tumour.	18 hours.		D.		D.	18 hours.

§ Turning ineffectually attempted. Craniotomy was also unsuccessfully performed.
‖ Dead before operation.

TABLE I. (*continued.*)

No.	Year.	Name and residence of the patient.	By whom and where the case is related.	Operator.
76	Sept. 10, 1864.	Mary Salimson, City of London Lying-in Hospital, aged 28.	Dr. Greenhalgh, private communication.	Dr. Greenhalgh.
77	Nov. 23, 1864.	Ann Burgess, No. 1, Brown Street, Acton Street, London Road, Manchester.	Dr. Thomas Pigg.	Dr. Clay.
*	Sept. 8, 1878.	M. S., aged 42, Manchester Workhouse Hospital.	Dr. W. F. O'Grady, private communication to the author.	Dr. W. F. O'Grady.

* The above case having come to hand after the Tables and deductions were already printed, it was only possible to give a bare statement of the particulars furnished.

TABLE I. (*continued.*)

Cause of difficulty.	Duration of the labour.	Mother.		Child.		Mother survived.
		Preserved.	Died.	Preserved.	Died.	
Distortion of the pelvis, from rickets.	About 41 hours.		D.	P.		48½ hours.
			D.			
Malacosteon.	Strong pains for 24 hours, but had probably been in labour to some extent before this.			P.		3 days.
Cancer of the cervix uteri.	2 hours.		D.		D. Premature, and dead before operation.	36 hours, died from exhaustion due to previous losses of blood.

TABLE II.

OF RECORDED CASES OF CÆSAREAN SECTION

No.	Year.	Name and residence of the patient.	By whom and where the case is related.	Operator.
78	Aug. 29, 1837.	Frances M., aged 38, Lambeth Workhouse.	Mr. T. Bryant, *Obs. Trans.*, vol. vi. p. 197 ; London.	The late Mr. T. E. Bryant.
79	April 27, 1863.	Mrs. S., aged 21 years.	Dr. M. T. Sadler, *Edin. Med. Jour.*, vol. x. p. 268 ; *Medical Times and Gazette*, vol. ii. p. 141, 1864.	Dr. M. T. Sadler.
80	July 31, 1864.	J. G., aged 31 years, Westminster Hospital.	Dr. Frederic Bird, private communication to the writer.	Dr. F. Bird.
81	January 1865.	A poor woman ; as she resided in a miserable room she was removed to the Lying-in Hospital, Belfast.	Dr. Pirrie, *British Med. Jour.*, vol. i. p. 94, 1865. Dr. Pirrie kindly furnished me with notes.	Dr. Pirrie.
82	1865.	...	Dr. Wiblin, *British Med. Jour.*, vol. xi. p. 261, 1865.	Dr. Wiblin.
83	Oct. 15, 1865.	Mrs. ——, an Irishwoman.	Sir J.Y.Simpson, Bart., *Edin. Med. Jour.*, vol. xi. p. 865.	Supposed Sir J. Y. Simpson, Bart.
84	Sept. 27, 1865.	Mrs. L., aged 28 years.	Dr. Greenhalgh, *Obs. Trans.*, vol. vii. p. 220; also an abstract of case, *Brit. Med. Jour.*, vol. ii. 1867, case vi., p. 400.	Dr. Greenhalgh.
85	Nov. 11, 1865.	Mrs. W., aged 32 years.	Dr. Greenhalgh, *Obs. Trans.*, vol. vii. p. 275; also an abstract of case, *British Med. Jour.*, vol. ii., 1867, case vii. p. 400.	Dr. Greenhalgh.

TABLE II.

IN GREAT BRITAIN AND IRELAND.

Cause of difficulty.	Duration of the labour.	Mother.		Child.		Mother survived.
		Preserved	Died.	Preserved	Died.	
Distorted pelvis, from rickets.	66 hours; 18 hours after rupture of membranes.		D.	P.		31 hours.
Supposed to be exostosis; found after death to be an enormous cyst filled with hydatids, which was rendered so hard as to represent a bony growth.	Some days in lingering labour; membranes ruptured 8 days before operation.		D.		D.*	Rather more than 24 hours.
Dwarf 3 feet 10 in. high; distorted pelvis; antero-posterior diameter, 1¼ inch.	72 hours before operation; 48 hours before the rupture of the membranes and 24 after.		D.		D.†	15 hours.
...	Had been four nights and days in labour before she came into the hospital.		D.	P.		About 20 hours.
Deformed pelvis.	...		D.	‡	‡	25 hours.
Deformed pelvis; found the promontory of sacrum projecting greatly forwards like the size of a closed fist.	Liquor amnii discharged, as far as I can calculate, about 96 hours before the operation; had slight pains several days; os uteri dilated to nearly full size many hours before operation.		D.		D.*	67 hours.
Distorted pelvis, from rickets.	About 28 hours; membranes ruptured and liquor amnii discharged 28 hours, or thereabouts.		D.		D.§	31 hours.
Distorted pelvis, from mollities ossium.	As far as I can compute, about 112 hours, or nearly 5 days; liquor amnii discharged about 24 hours.		D.		D.*	80 hours.

* Dead before operation.   † Head perforated before admission.
‡ No account.   § Dead before from craniotomy.

P 2

TABLE II. (*continued*).

No.	Year.	Name and residence of the patient.	By whom and where the case is related.	Operator.
86	1865.	E. B., aged 34 years, Middlesex Hospital, London.	Editor of *Lancet*, *Mirror of Practice of Med. and Surg.*, vol. ii. pp. 700 and 722, 1865, under the care of Dr. Hall Davies.	Mr. de Morgan.
87	Feb. 3, 1866.	M. G., aged 30, St. Bartholomew's Hospital.	Dr. Greenhalgh, *Brit. Med. Jour.*, vol. ii. 1867, case viii., p. 491.	Dr. Greenhalgh.
88	Jan. 23, 1866.	Ellen O., aged 27 years, Cottesmore, Rutland.	Dr. W. Newman, *Obs. Trans.*, vol. viii. p. 343, London.	Dr. W. Newman.
89	Mar. 29, 1866.	Sarah W., aged 37 years, St. Bartholomew's Hospital.	Dr. Richardson, *Med. Times and Gaz.*, vol. i. 1866, p. 362; Dr. Greenhalgh's patient, *Brit. Med. Jour.*, vol. ii. 1867, p. 491, case ix.	Dr. Greenhalgh.
90	May 27, 1866.	—— Barnes, aged 38 years, Westminster Hospital.	Dr. Frederic Bird, in a private communication to the writer.	Dr. F. Bird.
91	June 14, 1866.	A woman, a patient of the Lying-in Hospital, Liverpool.	Liverpool correspondent of the *Brit. Med. Jour.*, vol. i. 1866, p. 673.	Dr. Grimsdale.
92	Aug. 19, 1866.	Mrs. W., aged 29 years, Shoreham.	Dr. Greenhalgh, *Lancet*, vol. ii. p. 203, 1866; *Brit. Med. Jour.*, vol. ii. 1867, p. 491, case x.	Dr. Greenhalgh.
93	June 3, 1867.	Ann Kinsey, aged 21 years, Northern Etchells, Cheshire, brought to St. Mary's Hospital, Manchester.	Dr. D. L. Roberts. Read at the Obstetrical Society, Dec. 4, 1867. Abstract of Case, *Lancet*, vol. ii. p. 769, 1867.	Dr. D. L. Roberts.

TABLE II. (*continued*).

Cause of difficulty.	Duration of the labour.	Mother. Pre-served.	Died.	Child. Pre-served.	Died.	Mother survived.
A cancerous growth from os uteri and vagina. Strong bearing pains set in perhaps 2, 3, or more hours before its performance, but slighter uterine pain may have existed before.	The membranes were ruptured; liquor amnii discharged 10 days before operation.		D.	P.		About 40 hours.
A large epitheliomatous growth of the cervix uteri.	About 5 or 6 hours; membranes entire.		D.	P.		69 hours.
Extensive epithelioma of the cervix and lower part of the body of the uterus.	4 days; the liquor amnii discharged two days before the operation.	P.		P.*		Recovered.
Epithelioma of the cervix uteri.	Labour had not commenced, and the membranes had not ruptured.	P.		P.†		Recovered.
A large fibrous tumour impacted in the pelvis.	10 hours before the operation was performed, the liq. amnii having just escaped.		D.	P.		33 hours.
High distortion of the pelvis from disease of the spine and ankylosis of the right hip. The conjugate diameter did not exceed 1½ inch.	...	P.		P.		
A cancerous tumour of the rectum; solid and immovable; of such a size that in no part of the pelvis could more than the index finger be passed.	Not in labour; eight months and a week advanced in pregnancy.		D.	P.		6 days.
Deformity of the pelvis from an undeveloped state of the bones and malformation of the apertures.	The exact duration of the labour is a little uncertain; but as far as it can be computed, it was fully 48 hours. The liquor amnii had been discharged for a considerable time; it may be from 30 to 40 hours.		D.	P.		4 days 15 hours.

* Premature, 6½ to 7 months; born alive.　　† Lived for an hour.

TABLE II. (*continued*).

No.	Year.	Name and residence of the patient.	By whom and where the case is related.	Operator.
94	Oct. 6, 1867.	Martha Baggot, aged 21 years, 28 Stafford Street, Lisson Grove, British Lying-in Hospital, London.	Dr. Eastlake, *Brit. Med. Jour.*, vol. ii. p. 314, 1867. Further particulars, private communication made to the writer.	Dr. Eastlake.
95	Oct. 25, 1867.	Informed she was about 37 years of age; Guy's Hospital.	Dr. J. Braxton Hicks. *Obstet. Trans.*, vol. x. p. 45.	Dr. J. B. Hicks.
96	Dec. 28, 1867.	Mrs. H., 23 years of age, Pickering Place; removed to the London Surgical Home.	Dr. John Taylor, *Lancet*, vol. i. 1868, p. 85. Further particulars communicated to the writer.	Mr. Baker Brown.
97	Aug. 14, 1865.	A married woman, aged 24 years, Samaritan Hospital.	Mr. T. Spencer Wells, *Medical Times and Gazette*, vol. ii. 1865, p. 359.	Mr. Wells.
98	Feb. 25, 1865.	Mrs. H.	Mr. Walter Hardin, *Lancet*, vol. ii. 1865, p. 369.	Mr. Hardin.

TABLE II. (continued).

Cause of difficulty.	Duration of the labour.	Mother.		Child.		Mother survived.
		Preserved.	Died.	Preserved.	Died.	
Distorted pelvis, from rickets.	96 hours—viz., 48 before and 48 after, the rupture of the membranes.		D.	P.		4 days.
Distorted pelvis rostrated, from softening of the bones about puberty for eight months only.	Operation performed at 8½ months of pregnancy, inducing uterine pains by secale a few hours previously.		D.	P.		
Pelvis distorted, from an exostosis springing from the sacrum. The whole internal surface of this bone is thickened by bony deposit, which is greatly and suddenly increased at its promontory, reducing the ant. posterior diam. to 1¼ inch. The deformity is increased by the extreme angular position of the pelvis in relation to the spine.	18 hours; membranes not ruptured.	P.		P.		Lived.
No pelvic obstruction, but the operation was performed in consequence of a trocar having been thrust into the gravid uterus, after the removal of a large ovarian tumour, &c.	Labour did not exist; she was about five months pregnant.	P.			D. (When removed).	She lived.
Distortion of the pelvis from rickets. Brim, antero-posterior diameter, 1 inch; transverse, 4½ inches; oblique, 4¾ inches.	The liquor amnii had escaped six to seven hours. There were no uterine contractions.		D.	P.		4 days.

TABLE III.

OF RECORDED CASES OF CÆSAREAN SECTION

No.	Year.	Name and residence of the patient.	By whom and where the case is related.	Operator.
99	1856.	S. W., aged 39, Pendleton.	Mr. Henry J. Heywood. First related in this Table.	Mr. H. J. Heywood.
100	Aug. 1, 1856.	E. M., aged 40, Dromore, Ireland.	Mr. S. F. Hawthorne, *Medical Circular*, 1856; reported also to Dr. Radford by letter, July, 1880.	Mr. Hawthorne.
101	April 3, 1866.	Mrs. D., aged 30, Tanderagee, Armagh.	Dr. Crawford, *Medical Press and Circular*. vol. i. 1866, p. 624.	Dr. Crawford.
102	1867.	——, aged 35.	Fergus M. Brown, of Wausford. First related in this Table.	Dr. W. Steward.
103	1869.	Mrs. ——, aged 40, London.	Dr. Braxton Hicks, *Trans. Obst. Soc. Lond.*, vol. xi. p. 99.	Dr. Braxton Hicks.
104	July 29, 1869.	Mrs. B., aged 38, Alvechurch, Worcester.	Mr. J. S. Gaunt, *Brit. Med. Journ.*, vol. ii. p. 240, 1869.	Mr. Parsons.
105	Jan. 8, 1870.	H. S., aged 35, Ellesmere Union.	Dr. John Roe, *Lancet*, vol. ii. 1870, p. 149.	Dr. John Roe.
106	Aug. 26, 1870.	Mrs. L., aged 33, Glasgow.	Dr. A. Neilson, *Lancet*, vol. i. 1870, p. 335.	Dr. A. Neilson.
107	April 2, 1870.	Mrs. O., aged 35, London Hospital.	Mr. Stephen Mackenzie, *Brit. Med. Journ.*, vol. i. 1870, p. 409.	Dr. Head.

TABLE III.

IN GREAT BRITAIN AND IRELAND.

Cause of difficulty.	Duration of the labour.	Mother. Pre-served.	Mother. Died.	Child. Pre-served.	Child. Died.	Mother survived.
Large fibroid tumour, growing from in front of sacrum, and occupying entire inlet of pelvis.	36 hours.		D.	P.		7 days.
Malacosteon pelvis; nine previous labours, first seven were natural, eighth and ninth required craniotomy. Cæsarean section now performed.	26 hours.		D.	P. (And bore several children.)		3 days.
Malacosteon; the disease being rapid in its progress. The abdominal wound was made oblique, and was healed on fourth day.	Not mentioned.		D.	P.		6 days.
Malacosteon; patient in delicate health; first labour was a forceps case, and the next craniotomy. Wound healed and diarrhœa, apparently unconnected with operation, caused death.	24 hours.		D.		D.	14 days.
Transverse position, with fibroid tumour of uterus obstructing inlet of pelvis.	Many hours, and patient exhausted.		D.	P.		3¾ days.
Malacosteon; transverse diameter, 1 inch; ant. post. diam. 1¼ inch.	Two days.	P.		P.		
Malacosteon.	4 or 5 days.		D.	P.		4½ days.
Malacosteon; transverse diameter of outlet was 1 inch.	True labour, 4 or 5 hours.		D.	P.		14 hours.
Malacosteon; ant. post. diam. of inlet, 2½ inch; trans. diam. of outlet, ¾ inch.	Not mentioned.		D.	P.		15½ hours.

TABLE III. (continued).

No.	Year.	Name and residence of the patient.	By whom and where the case is related.	Operator.
108	Jan. 13, 1871.	M. A. S., aged 23, Wolverhampton Lying-in Ward.	Mr. Henry Gibbons, Trans. Obst. Soc. Lond., vol. xiii. p. 131.	Mr. Henry Gibbons.
109	1871.	Dwarf, age not stated, Kintore.	Dr. Andrew Inglis, Edin. Med. Journ., vol. xvii. p. 341.	Dr. A. Inglis.
110	Sept. 29, 1871.	M. A. M., aged 30, Romford.	Mr. Norris F. Davey, Lancet, vol. i. p. 826.	Mr. Norris F. Davey.
111	Mar. 17, 1872.	H. G., aged 22, Leeds.	Dr. Philip Foster, Lancet, vol. i. 1872, p. 753.	Dr. Philip Foster.
112	Sept. 15, 1872.	Maria N., aged 34, London.	Dr. G. E. Yarrow, Lancet, vol. ii. 1872, p. 523.	Dr. G. E. Yarrow.
113	May 5, 1873.	Mrs. N., aged 30, Wisbeach.	Mr. D. C. Nicholl, Lancet, vol. i. 1873.	Mr. D. C. Nicholl.
114	1873.	Mrs. Sykes, aged —, St. Mary's Hospital, Manchester.	Mr. Geo. W. Pettinger, now for first time. See sketch, brim of the pelvis. Fig 11.	Mr. Geo. W. Pettinger.
115	1873.	——, aged 28, St. Mary's Hospital, Manchester.	Dr. Lloyd Roberts. Now for first time.	Dr. Lloyd Roberts.
116	Jan. 6, 1874.	Dwarf, aged 21, St. Mary's Hospital, London.	Dr. Meadows, Brit. Med. Journ., vol. i. 1874, p. 106.	Dr. Meadows.
117	Apr. 29, 1875.	Jane B., aged 38, Queen Charlotte's Lying-in Hospital.	Dr. Grigg, Brit. Med. Journ., vol. i. 1875, p. 609.	Dr. Grigg.
118	July 17, 1875.	A Dwarf, aged 28, Birmingham.	Dr. Malins. Private letter to Dr. Radford.	Dr. Malins.
119	July 20, 1875.	Mrs. H., aged 32, Bury, Lancashire.	Mr. John Parks, Lancet, vol. i. 1876, p. 240.	Mr. John Parks.
120	Oct. 23, 1875.	E. P., aged 29, London.	Dr. Oswald, Trans. Obst. Soc. Lond., vol. xvii. p. 378.	Dr. Routh.
121	June 23, 1876.	K. T., aged 28, North Staffordshire Infirmary.	Dr. Walter, Brit. Med. Journ., vol. ii. Now related for the first time.	Mr. John Alcock.
122	Oct. 8, 1876.	M. H., aged 41, Guy's Hospital.	Dr. Galabin, Trans. Obst. Soc. Lond., vol. xviii. p. 252.	Dr. Galabin.
123	Oct. 8, 1876.	M. R., aged 28, Lond. Temp. Hospital.	Dr. James Edmunds, Lancet, vol. ii. 1876, p. 818.	Dr. Edmunds.

TABLE III. (*continued*).

Cause of difficulty.	Duration of the labour.	Mother.		Child.		Mother survived.
		Preserved	Died	Preserved	Died	
Rachitic pelvis; ant. post. diam., 1¾ inch.	5 to 7 hours.		D.	P.		40 hours.
Rachitic pelvis; ant. post. diam., 2½ inches.	Over two days.		D.		D.	Not a day.
Rachitic pelvis; ant. post. diam., 1½ inches.	23 hours.	P.			D.	
Rachitic pelvis; ant. post. diam., 1 inch.	Slight pains for two days.	P.			D.	
Rachitic pelvis; ant. post. diam., 1½ inch.	Over 24 hours.		D.	P.		4½ days.
Cancer of uterus and vagina.	Over 26 hours.		D.		D.	41 hours.
Malacosteon; patient in delicate health.	A long time.		D.	P.		6 days.
Cancer of uterus; patient exhausted.	12 hours.		D.	P.		24 hours.
Rachitic pelvis; ant. post. diam., 1⅝ inch.	Not stated.		D.	P.		32 hours.
Rachitic pelvis; ant. post. diam., 1⅝ inch.	6½ hours.		D.	P.		27 hours.
Most probably rachitic pelvis; ant. post. diam., 1 inch. It was stated to be cordate in shape.	6 or 7 days.		D.		D.	3 days.
Rachitic pelvis.	Early.	P.			D.	
Rachitic pelvis; ant. post. diam., 1¾ inch.	About 20 hours.		D.			62 hours.
Rachitic pelvis; ant. post. diam., 1½ inch; knee presentation.	24 hours, but not severe labour.	P.		P.		
Cancer of vagina and uterus.	6 or 7 days, much exhausted.		D.		D.	Died on the table.
Exostosis of ischium.	60 hours.	P.		P.		

TABLE III. (*continued*).

No.	Year.	Name and residence of the patient.	By whom and where the case is related.	Operator.
124	Mar. 13, 1877.	M. T., aged 28, Stanton Lees, Derbyshire.	Mr. E. M. Wrench, *Lancet*, vol. ii. 1878, p. 4.	Mr. E. M. Wrench.
125	July 26, 1877.	M. H., aged 84, St. Thomas's Hospital.	Dr. Jervis. First related in this Table.	Dr. Jervis.
126	Nov. 3, 1877.	M. A. S., aged 40, Middlesex Hospital.	Mr. Henry Morris, *Lancet*, vol. i. 1878, p. 488.	Mr. Henry Morris.
127	...	Patricroft.	Mr. Robinson. Related now for first time.	Mr. Robinson.
128	Feb. 12, 1878.	C. G., aged —, Guy's Hospital.	Dr. Braxton Hicks, *Trans. Obst. Soc. Lond.*, vol. xx. p. 106.	Dr. Braxton Hicks.
129	June 13, 1878.	Mrs. L., aged 23, Western Infirmary, Glasgow.	Dr. Hugh Miller. Now first related.	Professor M'Leod.
130		J. M'V., aged 28, Glasgow.	Dr. Hugh Miller. Now first related.	Professor Buchanan.
131	May 20, 1879.	——, aged 30, Stirling.	Mr. M'Nab. Now first related.	Dr. William Johnston.

TABLE III. (*continued*).

Cause of difficulty.	Duration of the labour.	Mother.		Child.		Mother survived.
		Preserved.	Died.	Preserved.	Died.	
Rachitic pelvis.	35 hours.	P.		P.		
Cancer of vagina and uterus.	About 1½ day.		D.	P.		2 days.
Malacosteon; ant. post. diam. available, 1⅝ inch.	32 hours.		D.		D.	60 hours.
Rachitic pelvis; version and craniotomy first performed.	...		D.		D.	
Cancer of vagina.	About 3 hours.		D.	P.		24 hours.
Ant. post. diam., 1¼ inch.; due to an injury to pelvis many years before.	31 hours.		D.		D.	A few hours.
Rachitic pelvis.	About 24 hours.		D.	P.		30 hours.
Cancer of uterus.	2 or 3 days.		D.	P.		8 days.

ON THE MEANS TO SUPERSEDE THE CÆSAREAN SECTION.

THE real maternal mortality after Cæsarean section, which, in my opinion, very considerably depends on the vacillating treatment of the practitioner himself in delaying the operation too long before it is performed, and has induced several writers to propose measures either to completely supersede the necessity of this operation, or to substitute another operation which is considered to be safer.

Dr. Blundell has proposed "Section of the Fallopian Tubes." He says : " Now is there any mode in which, when the obstruction of the pelvis is insuperable, the formation of a fœtus may be prevented ? In my opinion there is." To accomplish this, he advises an incision an inch in length in the linea alba above the symphysis pubis, and that " the Fallopian tubes on either side should be drawn up to this aperture," and " a portion of the tubes should be removed," which is "an operation easily performed, when the woman would for ever after be sterile."—(*Lectures*, Cassell's edit.)

I have never heard of a case in which this operation has been performed.

Dr. Blundell further makes the following remarks :— " Sixthly, *Extirpation of the Puerperal Uterus.*[*]—When the Cæsarean operation is performed, or when a patient is evidently sinking after rupture of the womb, let it be remembered that the wound formed by the extirpation of the womb, and which might probably be much reduced in extent by drawing the parts together with a ligature, would merely take place of a more formidable wound, that, I mean, formed in the womb by the Cæsarean operation, and which by the operation here performed would, together with the uterus, be taken completely out of the body."[†]

[*] Paper read before the Medico-Chirurgical Society, 1821.
[†] " Researches, Physiological and Pathological," p. 27, 1823.

The above quotation, with many others contained in Dr. Blundell's observations, point out most clearly that he ought to be strictly considered as the pioneer of abdominal surgery. Porro doubtless based his operation on Blundell's suggestion.

LAPARO-ELYTROTOMY.

In order to avoid the performance of Craniotomy with all its dangers to the mother, and also the maternal mortality of the Cæsarean section, the operation of Laparo-Elytrotomy was proposed as a substitute for Cæsarean section some years ago, but did not meet with much attention; although Dr. Thomas did not originate the sub-peritoneal section, yet he had the credit of opening the vagina by laceration to avoid hæmorrhage. There are doubtless certain contingent evils belonging to this operation as well as there are to the Cæsarean section. It is said, with a timely delivery, to have a mortality of 25 to 30 per cent. It is very unfair to make any comparison between the results of Cæsarean section and Laparo-Elytrotomy, as there is a considerable difference in both the character of the patients and their condition when the operations are undertaken. It has been said that this mode of delivery is infinitely safer than the Cæsarean section, but I do not believe that it will prove to be, as Dr. Harris says, any less dangerous than the Cæsarean section, performed, as it should be, early on the first day of labour.

There have been only two women in England upon whom Laparo-Elytrotomy has been performed.

PORRO'S OPERATION.

I HAVE alluded to this operation, and stated what the statistics were. It is an operation of great importance, and requires attentive consideration. I received in August, 1880, a paper extracted from the *American Journal of Medical Science* for July last on Gastro-Hysterectomy—a modification of

Dr. Porro—by Isaac Taylor, M.D., Emeritus Professor of Obstetrics, &c., New York. He says: "The last report of Dr. Harris on the Porro operation shows that up to that time (April, 1880) there had been thirty-six cases operated upon, of which one-half were saved and one-half died. By rejecting or excluding those cases which were unfavourable, owing to labour having existed for a long while, or through an exhausted condition of the system, the proportion would be materially changed, and would be as eighteen to eleven, nearly 62 per cent. saved."

"There are two modifications presented for acceptance to the profession, to which I purpose offering another. I will cite two of them :—

"1st. The Porro method :
　　1. Abdominal incision.
　　2. Opening the uterus, delivering the child and placenta.
　　3. Ligating the uterus by the serre-nœud and removing it.
　　4. Securing pedicle by clamp to the abdominal wound.
　　5. Drainage tubes through Douglas's space.

"3rd. Taylor, of New York :
　　1. Abdominal incision, four and a half to five inches.
　　2. Opening the uterus and delivering child only.
　　3. Ligating the pedicle with strong whipcord or fish-line as a temporary ligature.
　　4. Cobbler-stitch one inch below for permanent ligature.
　　5. Removing uterus with placenta by scissors or scalpel.
　　6. Dropping pedicle in the pelvic cavity ; no drainage tubes.

"The operation of removing the uterus is, from the standpoints I have taken, simple and easy ; in it there is as much of a pedicle to ligate and remove as in an operation for ovarian tumour, and it is as justifiable as the removal of any ovarian tumour. The pedicle is longer in some instances than in others, and is modified in structure according to the nature of the constitution of the female at the time of the operation—

thicker or thinner, denser or more friable. As to the ablation of the uterus unsexing the woman, it does no more than the Battey operation for the removal of the ovaries in the unimpregnated state. The future beneficial result, nevertheless, is to the woman of incalculable advantage and benefit. It claims an interest in her behalf in not allowing her to have any more progeny. It abrogates her right, justly, and, as I conceive, morally it should, to undergo another capital operation. It is a boon to her, though she may not be aware of the loss of the womb. It is absolutely necessary, in a social and political aspect; for women of that class who are deformed from disease, as is witnessed so often in Continental Europe, are not capable of giving that care and attention necessary for the sustenance of their children, who, as well as themselves, are more generally cast upon the public charities for support, and they virtually become, therefore, a lien on the finances of those charities. The operation may unsex the mothers so far as the loss of the uterine organs are involved, but it does not remove or extirpate the sensual enjoyment, as that remains as normal as ever. It is a moral obligation, I hold, therefore, of the highest duty to sacrifice that organ, as much so as it is held to be a high moral duty by some of the most eminent obstetricians, and none more so than by our celebrated Meigs, that the sacrificial act of the child should not be performed several times, and by some that not more than once in those cases of deformed pelves necessitating the performance of Cranioclasm, but to relegate it to the performance of the Cæsarean section, or the modification of that operation, or Laparo-elytrotomy. The success of the recent modifications of the Cæsarean section by Professor Porro is truly remarkable and very great. Only four years have elapsed, and there have been forty-nine operations."

THE END.

PRINTED BY BALLANTYNE AND HANSON
LONDON AND EDINBURGH